History of the Twenty-Second Infantry Regiment in World War II

History of the Twenty-Second Infantry Regiment in World War II

Compiled and edited by Dr. William S. Boice
Chaplain The Twenty-Second Infantry Regiment
August 1943–February 1946

Deeds Publishing | Athens

Published by Deeds Publishing in Athens, GA
www.deedspublishing.com

Printed in The United States of America

Cover and interior design by Deeds Publishing

ISBN 978-1-961505-24-7

Books are available in quantity for promotional or premium use. For information, email info@deedspublishing.com.

First Edition © 1959
Second Edition © 2024

10 9 8 7 6 5 4 3 2 1

To the officers and men who served honorably in the Regiment, and who gave both honor and life to the Regimental motto, "Deeds Not Words."

Contents

DEEDS NOT WORDS

TWENTY SECOND REGIMENT INFANTRY

The Twenty-Second Infantry Regiment Crest

The Regimental crest is composed of two main parts, a central shield embodied in an eagle and a smaller shield over the head of the eagle.

The main shield symbolized campaigns of five Indian Wars, 1868-90, with five arrows in the upper white portion of the field. Below this is symbolized the conquest of the Philippine Islands 1899-1906. This period is represented by the face of the Sun God of the Moro tribe imposed on the blue citadel of the islands.

The smaller shield with its Palm tree on a background of red and yellow of Spain, represents the invasion of Cuba, where the regiment was the first ashore; and the capture of Santiago in 1898.

Regimental motto —
"Deeds not Words"

This slogan is used to describe the unit's aggressiveness and discipline displayed while fighting the battle-hardened regulars of the British Army. Their gray uniforms made the British believe they were militiamen, but their skills in battle proved otherwise. The slogan was coined by a British General when asked by his adjutant if these soldiers were indeed militia during the Battle of Chippewa on July 5, 1815. His response was, "Those are Regulars, by God!" The battle signaled the point when the Regular Army gained the respect of its adversaries and renewed the American soldier's faith in himself.

Regimental slogan—
"Regulars by God"

This slogan is used to describe the unit's aggressiveness and discipline displayed while fighting the battle-hardened regulars of the British Army. Their uniforms made the British believe they were militiamen, but their skills in battle proved otherwise. The slogan was coined by a British General when asked by his adjutant if these soldiers were indeed militia during the Battle of Chipewa on July 5, 1815. The battle signaled the point when the Regular Army gained the respect of its adversaries and renewed the American soldier's faith in himself.

When Chaplain Bill Boice and Major General (Retired) John Ruggles turned over the leadership of the 22nd Infantry Regiment Society to me in August 1995, their instructions were to welcome all veterans who ever served at any time in the 22nd Infantry Regiment, and to preserve our history.

Bill gave me a copy of this book and said, "Maybe someday you can get this book professionally published."

At the time, I never dreamed I would become a publisher who specializes in military books and memoirs.

In this 80th anniversary year of the 22nd Infantry Regiment's fight across Europe from D-Day (6 June 1944) to VE Day (8 May 1945), I knew this is the perfect time to practice our Regimental motto – Deeds not Words – and publish this book for all to read.

Proceeds from the sale of this book will be used to write and publish more of the history of the 4th Infantry Division and the 22nd Infantry Regiment.

Deeds not Words! — Steadfast and Loyal...
Robert O. Babcock

Foreword

A History of the Twenty-Second United States Infantry Regiment prepared by Colonel John McAuley Palmer and Major William R. Smith, complete through May 1922 and published in that year is the only comprehensive history of the Regiment to that date.

This is the history of the Regiment in World War II, closing with the inactivation of the Regiment in 1946.

The history is based on the After-Action Report of the Regiment during combat and compiled by Lt. David R. James at Camp Butner, North Carolina, in 1945. Personal data and material have been supplied by Colonel Arthur S. Teague, Major General Charles T. Lanham, Captain Donald L. Faulkner, and from contemporary publications. Original material is supplied by the editor.

We readily acknowledge there were many important events and actions not adequately reported; such is the fortune of war. The material is presented as it has been remembered, and each man will remember events from his vantage point.

This volume is written to help us remember.

—**William S. Boice,** Chaplain
The Twenty-Second Infantry Regiment
4th Infantry Division 1943–1946

Glossary of Terms and Abbreviations

GI — Government Issue, i.e. "soldier"

IPW — Interrogation Prisoner of War

CP — Command Post

IP — Initial Point

ETD — Estimated Time of Departure

WP — White Phosphorus

FFI — French Underground Forces

I & R — Intelligence and Reconnaissance

SOP — Standard Operating Procedure

CT — Combat Team (usually a Regiment plus attached units)

Ml — Military Intelligence

OP — Observation Post

HMG — Heavy Machine Gun

LMG — Light Machine Gun

CO — Commanding Officer

LD — Line of Departure

FO — Forward Observer

HE — High Explosive

PW — Prisoner of War

AT — Anti-Tank

Bn — Battalion

Unit Strengths —
Normal Regimental strength = 3,200 men
Normal Battalion strength = 900 men
Normal Company strength = 240 men

1. The Years Between

Guarding the docks at Hoboken is not glamorous. Since this was the lot of the Regiment during World War I, its men could only be proud of its past record, its age, and its tradition, and gaze with jealous eyes at the ships and men who sailed for the shores of France.

In 1922, following its period of Hoboken guard duty, it was ordered to Ft. McPherson, Georgia. In 1927, the first battalion was inactivated and the third was ordered to Ft. Oglethorpe, Georgia. The third battalion, Headquarters Company, and Service Company were ordered to Ft. McClelland, Alabama, as their permanent station in 1935. For the next two years, the regiment shuttled among three stations, doing CMTC (Civilian Military Training Corps) and CCC (Civilian Conservation Corps) duty, along with American Red Cross work until 1936.

Until 1940, the regiment was a part of the old Eighth Brigade commanded by Brigadier General R. O. Van Horn. This was said to be the last infantry brigade in the United States Army, and it was not inactivated until General Van Horn's retirement in 1940.

In June of that year, the entire regiment was assembled at Ft. McClelland, with the exception of Company "F" which was left at Ft. McPherson. Colonel A. S. Peake commanded the regiment. He succeeded Simon Bolivar Buckner who commanded the regiment from 1939 to June 1940. Colonel Buckner had in turn succeeded Colonel John W. Lang.

Colonel Peake worked industriously to bring the regiment to a proper organization and high state of efficiency. Paul Turner commanded Company "F" during its status as a separate command, and in September 1940 the entire regiment assembled at Ft. Benning, Georgia. The men were moved into new buildings before there was heat, light, or water.

The regiment suffered its worst growing pains at this time. Recruits were assigned to the regiment for training and Major Raymond was put in command of the training unit. Machine gun companies had just been unhorsed and were becoming motorized in the army's attempt to come abreast of modern warfare.

With Colonel Peake commanding, Captain Paul Turner was adjutant, Major Raff was S-2, Major J. H. Halverson was S-3, and Lt. Colonel Almont Holly was S-4. Lt. Colonel Herbert Schmid commanded the first battalion and 1st. Lt. Earl (Lum) S. Edwards was his adjutant. Major Harvey J. Golightly commanded the second battalion with Captain Williams Stubbs as adjutant. Arthur S. Teague, a lieutenant in Company G, learned much about the army from Major Golightly. The third battalion command fell between Lt. Colonel Stewart Cutler and Lt. Colonel Holly.

In June 1940, Lt. Colonel George H. Weems took command. It was about this time the regiment was completely motorized. The end of June brought the famed Louisiana maneuvers from which the regiment returned to Ft. Benning in August, only to leave again for the Carolina maneuvers in October.

The attack on Pearl Harbor found the regiment ill equipped, but ready for duty. Immediately following a proclamation of war, they were dispersed throughout the state of Georgia guarding vital installations.

Training was continued at Camp Gordon, Georgia where the regiment had moved 20 December 1941, with the intensity born of war and the certain knowledge of eventual enemy contact. Gloom was further deepened by the fact that all personnel were restricted to the camp on Christmas Day. Highlight in the training of Gordon, long remembered, was the infamous twenty-five-mile road march in the summer of 1942. Specialized combat training was continued, and men were constantly cadred out from the regiment to newly forming units throughout 1942 and the early months of 1943.

In February 1943, Colonel Weems had left the command and was succeeded by Colonel Hervey A. Tribolet.

In April 1943, the Fourth Motorized Division, of which the regiment had become a part in 1940, was ordered to Ft. Dix, New Jersey, and another period of training was begun. In July, following extensive army experimentation, the division was de-motorized, becoming again the Fourth Infantry Division. This tour of duty was pleasant to the men of the Fourth because of its proximity to New York, Philadelphia, and Trenton.

After such cosmopolitan surroundings, the wilds of Carrabelle Beach, Camp Gordon Johnson, Florida, looked desolate indeed, but the regiment settled into specific amphibious training which was later to prove of much worth. Practice beachhead landings, swimming lessons, attacks on Dog Island, and the rigors of chilly November nights were all taken in stride.

While on a landing maneuver off Carrabelle Beach, Captain Clarence C. Hawkins, Regimental Motor Officer, was swept out to sea in a rubber life raft. No man to be thwarted by a mere ocean, "Hawk" sat tight knowing he would be missed. He very practically wrapped all his personal belongings, including his wristwatch, in a water-proof bag to prevent damage by the sea spray. Having been spotted by aerial observation, an LST reached Hawk and hove to and took him aboard. Still looking after his possessions, Hawk decided to put them aboard first — and with a mighty heave, threw them completely over the ship — and into the ocean beyond!

Probably no phase of the training of the regiment was more useful or more thoroughly detested than the time spent at Camp Gordon Johnson, Florida.

Schools were held for officers and non-commissioned officers. Boat loading tables were carefully studied, and boat assignments made. Personnel became familiar with such terms as LCI (Landing Craft Infantry), LCVP (Landing Craft Vehicular Personnel), and LCM (Landing Craft Mechanized).

On 1 December 1943, the Division proceeded by rail and motor convoy to Ft. Jackson, South Carolina for staging. Families lived in nearby Columbia in a sense of false gaiety over the Christmas holidays with the impending separation omnipresent. Days passed swiftly. Security measures were carefully observed; last minute checks were made on men and equipment.

On 6 January 1944, men took leave of their families and movement was begun to Camp Kilmer, New Jersey. The break had been made, and the division, anxious to get underway, had not long to wait. After thorough processing, the Twenty-Second Infantry Regiment loaded on board the British transport, "Capetown Castle," on the evening of 16 January 1944. In addition to the regiment, Division Headquarters and Company C, Fourth Medics were also aboard. Colonel Hervey A. Tribolet, Double Deucer Commanding Officer, was commander of troops.

Just before dawn on the morning of 17 January 1944, the

Capetown Castle slipped her moorings and put quietly to sea to keep her rendezvous with her convoy.

A North Atlantic crossing in January is never a good crossing, and January 1944 was no exception. It was cold, frequently rainy, and generally unpleasant. The convoy moved majestically into the broad expanse of the gray Atlantic. Since the ship was British, administration was typically so. Officers had comfortable staterooms and were served excellent meals in the dining salon. Quarters and rations for the enlisted men were inadequate.

Chaplains held nightly services in the mess halls, and in the face of the unknown, attendance was exceptionally good. Movies were operated by Special Services, and the ship's library was available for reading. There was time for sleeping, time for orientation of troops, and much time for wondering what was ahead.

Major John Dowdy was an officer who commanded the respect of the Double Deucers, but whose military bearing was no match for the rough Atlantic. Suffering from mal-de-mer, Major Dowdy took to his bunk, making the crossing for the most part horizontally.

A day or so out of Liverpool, he managed to make it to breakfast, one of the two meals served aboard ship daily. Looking peaked and pale, he kept control of the situation until the steward served another officer at the table his breakfast. When John saw the head and tail of a very British Kippered herring protruding from a bowl of warm milk, he turned green and left the presence of his fellow officers hurriedly, not to return that day.

The crossing took thirteen days, and on the afternoon of 29 January 1944, the Capetown Castle docked at Liverpool. Because of rail transportation difficulties, troops did not debark until the thirtieth, a gray, ugly day. This was a new world to the Twenty-Second Infantry. This was something to write home about. This was England.

2. Preparation For Invasion

England was beautiful, as beautiful and quaint as old fashioned, and sincere as another century. The Twenty-Second Infantry, de-training at Devon, was split and sent to the various camps which could accommodate it. Regimental Headquarters and the Second Battalion went to a camp outside the town of Denbury. The First Battalion was somewhat inadequately quartered in ancient and forbidding buildings in Newton Abbott. The Third Battalion, plus cannon and anti-tank companies, were stationed some distance away at South Brent in a camp which consisted almost exclusively of Quonset huts. Fourth Division Headquarters was at Tiverton. The nearest English city of any size was Exeter, a favorite shopping place soon to become known to us, and home of the beautiful and venerable Exeter Cathedral. To soldiers historically inclined, England was a source of much ancient tradition and history.

Torquay was the Atlantic City of South England and most of the officers and enlisted men went to Torquay for relaxation, usually to the excellent Red Cross Clubs.

Passes to London and surrounding cities of Bristol or Birmingham were difficult to obtain, for now the Division must settle down to stern and thorough training. Every precious moment must be utilized for the job which lay ahead. At a meeting of all of the officers of the Division, General Omar Bradley, then commanding the First Army chosen to storm the beaches, told the officers that originally it had been planned to storm the beach with one United States Infantry Division. This Division, picked by the top men of the General Staff, had been the Fourth Infantry Division. The officers returned to their regiments sobered, realizing that theirs was a job from which there was no turning back, and not many weeks later, this feeling of apprehension and determination was heightened by the statement of Major General Raymond O. Barton, Division Commander, that once we had hit the coast of France, we would have to fight to the last man, since before us would be the enemy and his fortifications and behind us nothing but open sea. There was little choice.

Training in squad problems, the handling of weapons, camouflage, use of artillery and mortars, assault tactics, pole charges, beehives, and the bazooka were given to the men, squad by squad, until they had become thoroughly familiar with their particular job. Work was blocked out. Certain tactics were taught, then company, battalion, regimental, and division problems involving these same tactics were run, in order to familiarize the troops with their practical application.

Weak spots within the organization were discovered and removed. Officers were shifted in their command. Day by day, the tension increased as it became evident that the long-promised second front would soon be a reality.

The Eighth Infantry Regiment plus the Third Battalion of the Twenty-Second Infantry Regiment were chosen to lead the assault on the beach of France, with the Twenty-Second Infantry and Twelfth Infantry Regiments following immediately. For this reason, the Third Battalion, with the Eighth Infantry, was sent to Widdecomb for special assault training, a training which paid for itself many times over in the course of the war for the specialized knowledge which it gave to its men.

Thus, the weeks turned into months and May arrived, bringing with it the tactics which involved the actual loading of the combat troops on the landing craft and running an assault landing at Slapton Sands. Causeways had been constructed with an inundated area, with impact areas for the artillery, and an objective 35 miles inland on the English Moors. After the invasion occurred, it seemed incredible to us that German intelligence and common sense could have missed the implications of the inundated area and the entire layout of these problems.

It was with mingled feelings that the elements of the regimental combat team received orders to move to the marshalling area.

These had been interesting days, charged with tension and anticipation, days when we had come to know and respect the British, and to appreciate this brave island.

All equipment was carefully packed and waterproofed. Personal effects, save only that which could be carried, were stored in foot lockers, and sent to the Effects Quartermaster in Liverpool for indefinite storage. All documents which might identify the unit were destroyed. No identification other than dog tags or AGO was retained.

On the night of 18 May 1944, the regiment moved by foot under cover of darkness to an assembly area. Then, shivering and cold, the men were moved by truck to three marshalling areas designated as X, Y, and Z, well away from towns and villages on the southern coast of England. Civilians were kept from the area or were required to remain in the area until after the invasion had occurred. The preparation and secrecy surrounding the pre-invasion were of the highest order. While aware of the impending action, the officers and men of the regiment operated on daily stand-by orders without specific knowledge of invasion plans.

An era had passed; preparations were finished. Mail home was curtailed and impounded, heightening the thoughts of loved ones and homes.

Each soldier wondered about his own courage. Yet, there was an eagerness to get into battle, to join the action against the enemy, and there was little expressed fear. It was not even possible to express personal uncertainty in letters; thus, the pressure began inexorably to build.

Last-minute equipment was requisitioned and obtained. Quick courses in elementary French, designed to give each man enough knowledge to ask for help in case he was wounded or food if he was hungry, were taught by qualified instructors. There was much joking about the lack of the French phrases most of the men were interested in.

It was while in the marshalling area that the new type demolition, the Beehive, was demonstrated to the assault companies. This new employment of a long-known principle of physics pleased GI's who soon caught on to its use and wanted to demonstrate more times than there were Beehives to be used for demonstrative purposes.

Chaplains conducted worship services daily. A regiment that had given little indication of religious interest prior to this time now began to take its religion seriously.

Jewish officers remained on duty while Catholic and Protestant men were at divine services. The Jewish Chaplain from Division Headquarters conducted Jewish services for the men of his faith. Attendance increased so sharply as to warn the Chaplains of the magnitude of their spiritual responsibility.

Chaplain William K. Hogg had been assigned to the Third

Battalion and very quickly assumed a place of respect and spiritual strength among his men. Throughout the campaign in Europe, Bill Hogg was a never-failing tower of strength for his men, tireless in his service, and effective in his ministry.

Chaplain Arthur J. O'Leary was the senior Chaplain of the regiment, and the Regimental Chaplain at this time. Father O'Leary was the oldest Chaplain both in years and in time of service in the regiment. He was direct, commanded the respect of his men and his chaplains, and he never compromised with what he felt to be right. He was a familiar sight with his assistant, Joe Melis, holding mass in a field, or exchanging Irish humor with some friend.

Chaplain William S. Boice was assigned to the Second Battalion.

The strength and service of the Chaplains of the Twenty-Second Infantry Regiment can never be measured. It is enough to say that during these crucial times, officers and men realized as never before the worth of their Chaplains and the seriousness of the task ahead. Spiritual strength counted.

Elements of the Fifth Armored Division ran the marshalling guards and did everything possible to make these last days in Britain as pleasant and comfortable as possible for every member of the Division. By this time, the men knew the impending assault was but a few days away. One soldier remarked that he felt like a turkey prior to Thanksgiving. The remark was greeted with laughter at the time; it ceased to be funny within a matter of days. All mail was impounded, both outgoing and incoming. Sand tables were constructed to an exact scale of the beach which was to be assaulted. The exact location of the beach, however, was still carefully kept secret, and only the battalion staffs, plus the company commanders involved, knew the place of the assault. Not even they had been told the day or the hour. The sand tables were placed in tents and carefully guarded as the men were brought into the tents in small groups. They became familiar with the terrain so that they could recognize physical features, either by silhouette or by sight.

The time in the marshalling area seemed incredibly long to soldiers who were keyed to a pitch. At last, orders came to move out and again, under the deep cover of darkness, the companies loaded in trucks and moved slowly over the steep hills of England toward the coast. The trucks halted on back country roads at about 2:00 A. M. and the soldiers, loaded with assault equipment, sat unbelievably cold in the sharp English night. No movement was allowed and there was nothing the weary soldiers could do but miserably await the morning. The night was clear and there was no fog. Lights were prohibited and smoking was allowed only within the covered trucks. A trucking unit provided the soldiers with breakfast put up in paper bags, and hot coffee was supposed to be served. The coffee was cold and bitter. The breakfast consisted of thick, dry slices of English whole wheat bread with cold bologna and a cold hard fried egg. Among other marks of war, surely one part of the English countryside was covered with the brown stains of disdained coffee and the wreckage of the world's worst breakfast.

Prior to dawn, the soldiers dismounted and already weary under their heavy assault loads, they moved up the steep hill and toward the loading harbors. It was with relief, not unmixed with anticipation, that they saw the countless landing craft and assault boats that were to convoy them across the channel. It was well past midday when the men got aboard and moved to their cramped quarters below deck. Space was at a premium; officers and men alike crawled into web-laced horizontal bunks barely far enough apart for both the man and his equipment. No men were allowed on deck. Time was spent in final briefing, and impregnating clothing with an evil-smelling preparation to resist gas. Men were warned to expect gas attacks on the beach.

Meals were a problem, and the men became acquainted with the 10-in-1 ration. Better than the K ration, and with a greater variety, the cartons contained canned foods, soluble powdered drink, crackers to take the place of bread, jelly or jam, canned milk, and toilet paper. Of all the ignominy of war, nothing so irritated and amused the men as khaki colored toilet paper. Surely, they felt, this was the last ignominious word in modern camouflage. The ration contained enough food for ten men and thus was called 10-in-1.

During the days of June 4 and 5, the men were given their silk escape map of France and a small compass. Briefings were stepped up, and each man's boat position and duty were carefully reviewed. The assault teams had been broken up into boat teams since the men would have to transfer from the larger craft to LCI's capable of getting close to the beach. Platoon sergeants were told what to do in the event a part of their team was lost or sunk.

Tension was high. Battle was ahead, but individual reaction was unknown. Most men worried more about their friends than about themselves. Esprit de corps was never better in a fighting unit, and whatever the personal fears the night of 5 June, collectively the men knew they were a great team, and they were confident. The men drifted to fitful slumber, lulled by the steady throbbing of the ship's engines. Their date with destiny would take place on the morrow.

The leaders of Combat Team 22 pictured in the marshalling area just before loading for the assault on D-Day. Seated, left to right: Colonel H.A. Tribolet, Commanding Officer 22nd Infantry; Lt Colonel John F. Ruggles, Regimental Executive Officer 22nd Infantry; Lt Colonel Arthur Teague, Commanding Officer 3rd BN 22nd Infantry; Lt Colonel S.W. Brumby, Commanding Officer 1st BN 22nd Infantry; Standing: Lt Colonel William A. Atson, Commanding Officer 44th Field Artillery; Lt Colonel Thomas Kenan, Regimental S-3 22nd Infantry; Lt Colonel Earl Edwards, Commanding Officer 2nd BN 22nd Infantry—Photo from the 22nd Infantry Regiment yearbook published in 1946

3. We Crack the Fortress, June 6-19

"No troops become seasoned, battle-hardened troops until they have fought their first battle. The invasion of France must be a success. We must therefore fight our second battle first."

—*Major General R. O. Barton*

As dawn approached, our invasion teams were approximately eleven miles off the coast of the Cherbourg peninsula. All night, planes had been roaring overhead, bombing the coastal defenses, or canalizing the beach obstacles. At 0100 hours June 6th, 1944, paratroopers had been dropped at the base of Cherbourg peninsula to seal off the eastern half of the peninsula. Previously it had been planned that the airborne troops would entirely seal off the peninsula at the base, but unexpectedly a German armored unit had pulled up into the western half, making this impractical.

Shortly thereafter, the fleets of both the United States and Great Britain commenced the shelling of the coastline and the beach defenses. Naval guns fired continually in an effort to soften the beach defenses for the infantry which was soon to arrive. The Third Battalion had two destroyers attached for direct support. In order that this fire could be coordinated, a shore fire control system had been organized. This shore fire control party consisted of an artillery officer and a naval officer, each with a radio party.

Although the waves broke high, the transfer was made from LCVP's to the assault boats in the crisp dawn without incident. The boats, difficult to control, bobbed about like top-heavy corks. They were crowded with men crouched in position, straining to see what the distant French coast looked like. Suddenly it appeared, the coast of France, lying low in the water and outlined by the intermittent flashes of gunfire from the supporting battleships, cruisers, and destroyers. This barrage was followed by the rocket boats firing salvo after salvo on the beaches and underwater obstacles. The noise, though great, was still distant enough for the men to have a detached feeling of unreality about the lethal effect.

The coastline seemed to come closer, but it was only that dawn was turning into day and the outlines of the Naval task force could be seen as it continued to fire on beach fortifications. Suddenly the motor of the assault crafts came to life and the regiment was headed for Utah beach. Equipment was slung loosely over one shoulder, for if a mine was struck or the boat was hit by a shell, heavy ammunition or equipment might drag the men under. The boats surged through the water; hearts were in throats and breath came quickly. The shore seemed miles away and time stood still; it was a state of suspended animation. Gigantic action unfolded slowly before our eyes. And then, the spell was broken, for the boat commander gave the order, "Get down; brace yourselves; prepare to land." There was a scraping sound as the craft hit a sandbar but shook herself free for another hundred yards. The ramp dropped and we stormed the beach.

How good to be ashore, to be free from the helplessness and insecurity of the water. It was very like amphibious training at Carrabelle Beach and Slapton Sands, but with one difference. Before us lay the first American soldier killed on the beach. His face was shattered and bloody and his body strangely twisted. Men looked at him curiously, then averted their faces as they walked by. Thus, death came and was to become a feared and tragic companion in the days ahead.

The Combat Command was brought ashore by Colonel Hervey A. Tribolet, a man for whom the regiment had the highest respect. His training of the regiment proved to be thorough and effective in the fighting days ahead. "Trib" was a fine soldier and a respected and loved Commanding Officer who knew his men and who led them well.

4. Personal Narrative

The Third Battalion, commanded by Lt. Colonel Arthur S. Teague, made the initial assault on Utah Beach, attached initially to the Eighth Infantry Regiment. The following narration is a verbatim report by Lt. Colonel Arthur Teague on the assault landing of the Third Battalion.

NARRATION — BY LT. COLONEL ARTHUR S. TEAGUE, JUNE 6-8, 1944

From landing craft, we came ashore on LCM's (Landing Craft Mechanized) — three of them — operated by Navy enlisted men. The Navy enlisted man on our LCM remarked that this was the third landing in which he had participated and that he didn't mind the initial landing so much as he did the ones afterwards because he would have to keep bringing in supplies.

Just as we were coming into the shore, I saw a shell that was fired from up the beach, and I knew some of us were going to be hit. I could see the spurts of water coming up. I saw one small landing craft hit, and thinking the same might happen to us, I told the Navy man to ram the beach as hard as possible. He said he would, and after holding it wide open for about two hundred yards, we hit the beach and stepped off on dry soil. A couple of boats behind us—about seventy-five yards back in the water—were hit, and then I saw a number of casualties. Many were killed and quite a few wounded.

I started up by the sea wall on the sand dunes and stopped for a moment. It was then that I heard someone call me. It was General Roosevelt. He called me over and told me we had landed 'way to the left' of where we were supposed to have landed, and that he wanted us to get this part of the beach cleared as soon as possible. He wanted action from my men immediately after landing and asked me to get them down the beach as soon as I could. This was about 0930.

At this time, we were getting quite a bit of artillery fire from the inland side of the beach. It was not very heavy, but spasmodic. I went on over and called a couple of officers on the staff and got behind the sea wall and suggested that we figure out what we had to do. We talked it over and thought about what could happen and decided the best thing to do was to find Captain Samuels, the Company Commander, and see what troops were already on the beach so that we could take stock of them.

A couple of tanks were on the beach. I yelled to one and crawled up on it. I asked the enlisted men about firing on the beach on the troops we could see. He stated that he had strict orders to just sit there and protect the troops coming ashore, and that was all. I told him, "For God's sake, start fire so we could reduce the troops waiting for us." He said he had orders to defend until the troops went through.

We started up the beach and I hollered back to everybody and got them dissembled because I saw two men who were lost on mines. I stayed on the sand dunes to see if I could identify my location on the map. Standing with my back to the water, looking inland, a little bit to my right front was the little round windmill or silo standing up which I had observed on aerial photographs and panoramic views of the beach before, which gave me the immediate location of where we were. I tried to get higher on the sand dunes, but someone yelled at me that snipers were firing and for me to get down.

I started on up the beach wall and ran into more troops. They said Lt. Tolles had been shot. On my way there, I passed along a number of baby German tanks which had electrical wiring and were loaded with TNT. Some troops wanted to fire into one and I told them to stop that action, and I posted guards on it. I went on around this little firing trench marked by barbed wire and sandy beach grass. Near this firing trench I went behind a sand dune into an open place and found Lt. Tolles lying on his side near another wounded man. I asked him what happened. He said he saw a white flag and he tried to get them to surrender, and someone had fired on him.

I immediately sent someone back to notify a doctor to move him out of the place. I went further up and ran into members of his platoon who had stopped and were having quite a little rifle fire back and forth. I saw what was happening as they moved along. My German interpreter was with me. We ran and hollered to them, and he yelled to the enemy in German. I ran on top of the sand dune. There I picked up an M-1 rifle and called to our men to get going. We went forward and suddenly encountered direct fire. I saw two Germans wounded. About seventeen of them raised up from different places around and started running across the beach. Pvt. Meis yelled at them in German. I questioned them and asked them where their mines were and about the number of Germans. They said they didn't know—that they had come only the night before. I told them they did know and that they would go with us.

I then started a skirmish line up the beach. They went about fifty yards up the beach and yelled, "Minen!" They started showing paths we could take to get out of there. I had seen Lt. Burton and Sgt. McGee wounded by mines along the beach. We moved on down the beach and picked up about 40 more Germans. Where they came from, I do not know; evidently troops ran them out. They came with their hands up and ran down the beach. We got on up a little farther and ran into a steel gate which I thought was T-7 entrance but now believe it to have been an entrance to U-5 causeway. I got hold of Lt. Ramano, Engineer Platoon Leader, and told him to open up the gate and while he was doing it, to have his engineers go up ahead and to lift out any mines.

I had gone up the beach a little farther and heard that my tanks were ashore, so I sent someone down there to get ahold of the tanks and to tell them to come on down the beach. This A Platoon, under command of a lieutenant from Alabama—I've forgotten his name—came up the beach about this time and we ran across from the little fortification on the beach wall. The Germans were firing down the beach a little and I could see these shots were hitting in the water. Some skimmed the tops of our heads, and some hit small boats. One of our tanks came up and got fired on and hit by small caliber guns. It was then that we noticed a small steel turret mounted on top of a pillbox and was moving along behind the beach wall. Our tank was about twenty-five yards away, but it immediately elevated its guns and opened fire, knocking the turret completely off the little fortification. Here we got quite a few more prisoners.

In the meantime, our men were having a pretty good fight inland near an old French fort where they had taken about a hundred prisoners. As we pushed on up the beach, our tanks were firing along the whole time. We found another steel gate of the Belgian type near the beach. It had been used quite a bit by vehicles before we landed. I positively identified it myself as being near T-7. I told Lt. Manor to get that out of the way. I had a tank. I pointed the gates out and he opened that entrance. I waited until he finished the job.

I continued on up the beach right in behind several units of our company and ran into Captain Samuels. Captain Samuels talked about one of the little tanks which had pushed around the entrance to T-7 and had stopped and been fired upon about three times by guns. The shots ricocheted off the tank and the Lieutenant fired the first shot, which went through the pillbox, which was the fortification we were supposed to have landed in front of. About twenty-five Germans ran across the beach with their hands up.

The companies pushed on to the fortification, and there I was with Captain Samuels, Captain Walker, and almost all the battalion staff. Major Goforth joined us and had I Company to hold up this point and L Company to attack normal buildings and the entrance to Causeway S-9. The attack was supplementary. At the time, we were getting mortar fire, so we three officers, plus Pvt. Buchavellis, decided we would dig into the sand dunes on Tare Green Beach. We dug about two feet into the sand and finally I remarked that that wasn't going to do any good because we weren't getting any of the other fortifications.

We kept noticing the gunfire that was coming down the beach so I took the platoon leader, and he and I crawled down the beach to see if we could observe where they were firing from. While we were lying there, the Germans saw us and fired two shots. One went over our heads and hit the water. The next one ricocheted off the tank which was close to us. We called for another tank. Firing continued from the S-9 fortification, causing quite a few casualties. Our tank fired a few rounds at it and finally destroyed it.

The mortar fire had let up a little by this time, which had been coming down from up the beach. I had just learned that one of our men with a flame thrower ran about twenty-five Germans out of a pillbox. He had taken two American paratroopers from that same pillbox.

I started out from this fortification straight across the mine field. I saw a house on fire. Behind me was Captain Walker and Captain Williams and quite a string of men. As we walked across this area, which had been dry at the time the mines had been placed in the ground, we could see several places which we knew mines were, because we could see where rocks had been pried up. I took out some white engineer's tape which we all carried, and we marked them as we went. I told them to step in the same tracks that I had made. As we walked, I heard one explode behind me. Captain Williams hit it and he got it through the cheek of the buttocks.

We went on across the mine field and found L Company. Here we met Captain Blazzard, who had machine guns set up and had been firing. I ordered them to assault the house and the S-9 nest simultaneously. This was a matter of about thirty minutes. I yelled for Captain Ernest to get him to hold L Company because I wanted to send K Company into attack.

All this time there was a gun still firing up the beach. It later developed that we could see where two or three shots hit the embrasures, but the Germans had destroyed it themselves.

About this time, I told Captain Ernest we could make an attack on the water's edge. We went out on the S-9 fortification about two hundred yards. The roads seemed to be in excellent shape, showing they had been used. We found a French civilian in one of the houses, so we asked him where the mines were. He pointed out that the road from S-9 up the beach was mined. In fact, he showed me about eight or ten mines. You could see where the mines had been put under the rocks. He said that the road hadn't been used for about four months. He said the other road was being used, and, to the best of his knowledge, was not mined. We pushed around for a short time and K Company jumped off and made a flank attack. I went with a battalion staff behind K Company. I started wading in water up to my waist, and in some places, up to my armpits. A long column of men was wading through the water. A sniper got a man just ahead of me. He lay there for most of the whole night because he couldn't be evacuated.

I followed K Company on up and encountered Lt. Pruzinski. He talked to Captain Ernest and told him that there was supposed to be a flame thrower behind the house, so I sent the Lieutenant out.

Then we went on up the beach and hit the causeway. We were getting quite a bit of fire and also quite a bit of mortar. Finally, K Company was able to take the approach to the causeway. Lt. Pruzinski had two tanks and he captured that point.

K Company cleared out the causeway and a few buildings at the end of it, and as it got late at night, I told Captain Ernest that we couldn't make much more distance, and we made preparations for the night.

There was a house there which we were afraid might be a booby trap. The men began digging into the place, but it was flooded with water. We were getting machine gun fire from the fortification ahead of us, so I told Captain Ernest that since we couldn't dig in, we would sleep along the road, and I would stay with the group. We lay down sometime around 12:30 at night, although it was hardly dark. We stayed there for the night. Captain Ernest, Captain Walker, and Major Goforth were with me. I told Ernest to tell the men we could sleep there tonight and that we weren't going to give up an inch of ground.

We put two machine guns on the causeway, and there was water all around us. It was about 1:00 A. M. before all was quiet. Then we began to make plans for an attack at 4:30. We worked out the plans on the map.

We continued the K Company attack the next day. We had the engineer platoon start removing mines from S-9 along the beach road. He worked all night. A machine gun kept him from removing them as fast as he could have otherwise. He had to work on his stomach all the while, but before daylight, he got the road pretty well cleared. After daylight he had all the mines out.

Two 57mm. guns were brought down the road from a house to the front lines to the little embankment which we had slept behind. All during the night a machine gun had been firing at the embankment, about two feet over our heads. There were about two hundred and fifty men along that road during the night. We got these 57's up, and I took Lt. Etta and showed him where the two guns were to go — one on the causeway and one behind the embankment. I pointed out the fortifications and told him I wanted the guns to be able to fire on them direct. I also got a tank. The larger guns had been knocked out during the night.

Here we tried to make an attack on them the next morning. We got off about 9:00 A. M. K Company tried to make a flanking attack sometime during the morning. It went through the water and set up a platoon. They were up to their necks in water. They were slaughtered in the water by machine gun fire. Captain Ernest said something had to be done about it. He grabbed a patrol and jumped into the water and yelled at them. He actually took the fire of machine guns from these men, because the Germans fired on him instead.

I ran down the road toward the 57mm. gun. It had ceased firing. Sgt. Thomas was behind the gun. I stuck one or two rounds in the 57 and let go with it. As soon as I fired, back came machine gun fire. Then we got some smoke from 4.2 mortars from Captain Williams and got K Company out of the water—what was left of K Company.

By that time, we had cleaned out two or three houses on the beach. It was approximately forty yards of dry beach. We got two machine guns into the houses. They began firing on the fortification about three hundred yards away. I sent a tank up the beach wall and got the bridge reinforced. We did everything possible to get the fortification to surrender, but it did not.

We fought a good part of the day, and in the afternoon when we had practically given up getting it to surrender, there was a fortification near Ravenoville where the Navy claimed they had seen a couple of white flags. We got permission from the regiment, left one company, about half of the mortars, and made a flanking attack with I and L Companies. We went out on the beach and started to Ravenoville.

Coming off this area from the water side from our position there, we had captured about twenty prisoners. Pvt. Meis, in talking to the German staff sergeant and private, found out that they had come from the fortification, which was the one we wanted to take. He stated that some men and two officers had been killed and that they would surrender if we could get to them, provided that one of the officers hadn't taken command. They further said that when the men wanted to surrender the fortification earlier that day and had tried to put up white flags, that the officers had fired on them and that they had fired back.

We kept this German sergeant and private and made the flanking attack about two miles down the road. Going down the road together were Captain Gatto, Captain Walker, and myself. It was about dusk when we got there. We decided we would send this German private in. We went further and saw a mob of men and so we dropped some smoke and he marched in. About eighty enemy surrendered at this fortification. We got them lined up and singled the one out who knew about mines on the beaches, another who knew about fortifications, and still another who knew about supplies. We left a medic to take care of the wounded. We marched the other men to the Regimental Command Post.

That night, we had the engineer platoon come in and put in a one span bridge over a bomb crater, which had been blown up so that water would flow across the road. During the night we got tanks to come down to our place on the beach. Staying with me that night were Captain Bridgeman, Captain Gatto, Captain Walker, and Captain Huck.

K Company was on the opposite side from us, about a mile away. In between us we had this German fortification from which we had captured prisoners, but which did not surrender. We slept in a blown-up place on the beach wall.

During the night, our C-47's were bringing gliders in. Ack-ack went up from the fortification. We fired mortars and silenced them from firing the ack-ack. Next morning, we were making plans to assault the place from both sides of the beach. We were ready to begin the assault when I was ordered to report to another place to help ward off an attack. Arrangements were made that the engineers would blow up the pillboxes and houses full of Germans. There were about twenty-five houses there. This was off the causeway from Ravenoville.

I started out with the company in formation. I got a few men across the causeway and this fortification opened up with machine guns and fired 20mm ack-ack also. We had some casualties. Our machine guns fired at them, but we couldn't get it stopped. I jumped on the side of the platoon sergeant's tank of the 776th Battalion, and told him I was going on the causeway, and I went and lay down and observed where the machine gun fire was coming from. I told him to come along beside me in the tank and adjust his firing. He did so and they directed a great deal of fire. It was hit on all sides. We got off about eight or ten shots from the tank and hit the back door of the fortification. We tried to shoot the entrance. About fifteen Germans ran out and across the field but were stopped after about fifty yards when the tank fired two rounds at them.

Then a fortification which was so well camouflaged that we hadn't seen it began to fire. We changed positions and fired at the second fortification. We got off about ten more rounds before they ceased their fire.

I had the tank placed so it could catch any fire, and after I got the men across, I jumped on the tank, and we got through okay. Going out we stopped and fired at pillboxes alongside the road.

5. The Critical Twenty-One Days

The plan for the invasion of France was both practical and successful. It required a short supply line in order to provide strength in depth for the invasion forces. It required an acceptable port for shipping to be made available to the Allies in the shortest possible time.

For this reason, it was the intent of the Supreme Allied Commander to put the invasion forces ashore on Utah Beach at the very base of the Cotentin Peninsula.

The airborne drop of the 101st and 82nd Airborne Divisions was intended to seal the base of the peninsula to prevent German reinforcement. With these divisions holding the base, the mission of the Twenty-Second Infantry Regiment was to turn right from the beach, storming the fortifications of Fortress Europa north toward Cherbourg. The intent was to secure the port of Cherbourg for allied use in a minimum time.

The Third Battalion of the Twenty-Second Infantry Regiment was attached initially to the Eighth Infantry Regiment, with the mission of crossing the beach seawall, turning northwest, and reducing the fortified positions along the coast. The Third Battalion, commanded by Lt. Colonel Arthur S. Teague, was the first element of the regiment to land, at approximately 0745 hours. The landing by the Third Battalion was made on Utah Beach with two assault units, Companies I and L, plus the Battalion Commander's free boat. Company I, directed by Captain Joseph Samuels, swung to the right 2, 000 yards below the proposed landing area and went into the attack. This was fortunate for the battalion, because at this point the beach seawall suddenly curved inward, and the German artillery pieces, sighted to fire down the seawall, were unable to traverse far enough inward to fire on this area.

Company L moved into an assembly area to the rear of Company I. Fifteen minutes later, Company K and Battalion Headquarters landed, and went into an assembly area just over the dune line. This battalion, with its tank support, immediately started down the beach reducing the pillboxes and strongpoints. These beach fortifications were actually an important feature of the Fortress Europa, and in order for any operation to be carried out on the Cherbourg peninsula, they must first be destroyed. Only two months before, these very fortifications had been visited by Hitler who had openly boasted that "Fortress Europa" was impregnable. The positions were mutually supporting along the narrow strip of land between the high-water mark and the inundation. The area was heavily mined.

The Third Battalion had Naval gun fire and 4.2 mortars, utilized to replace temporarily the artillery fire, and the successive enemy positions were bombarded by the Navy before being assaulted. Then tanks from a platoon of Company A, 746th Tank Battalion were brought up for point-blank fire while the infantry maneuvered inland from the rear of the pillbox. At 1030 hours, the 44th Field Artillery Battalion, in support of the Twenty-Second Infantry Regiment, landed. At 1038 hours, Battery C was firing in support of the Third Battalion. This battery was the first American artillery to fire on the beach fortifications from positions on Utah Beach.

The assault troops were subjected to heavy artillery fire from enemy forts farther to the rear. This fire was directed by the Germans on the positions being assaulted. Information was furnished by means of underground cables from each of the forward positions to those in the rear, resulting in extremely accurate artillery fire. At one point along the beach, the front was so compressed by the flooded area that the battalion was forced to stage a holding attack on the fortification defending that area, and to envelope the town of Focarville from the left (west), thereby reducing the effectiveness of the fortification. In order to accomplish this, the Battalion moved across a causeway into the First Battalion area and attacked, securing Focarville by dark.

The First Battalion, commanded by Lt. Colonel Sewell M. Brumby, and the Second, commanded by Major Earl W. Edwards, landed at H plus 210 in LCI's behind the Third Battalion, Twenty-Second Infantry, and the Third Battalion of the Eighth Infantry. Their mission was to cross the beach and the inundated area, attack to the northwest, reduce the strongpoints at Crisbecq and Azeville, and then secure the high ground west and southwest of Quineville. The inundation had been accomplished by the flooding or complete saturation of the soil for a width of approximately two miles inland, resulting in wide, marshy areas which were difficult to bridge and through which travel was restricted to established routes (causeways). Their landing was also made approximately 2,000 yards south of the proposed beach. Temporary confusion arose from the mistake, but the two battalion commanders soon organized their forces and attacked. They encountered much more resistance than was expected and did not reach their objectives. By nightfall, the Regiment was short of its initial objective but had reached the general line Utah Beach-Focarville.

The next morning, June 7th, came and despite the falling rain, the troops slept. They had fought doggedly until very late that initial day. At the first light of dawn, it was possible to see the non-coms and officers organizing their units for the oncoming attack. There was a tenseness in the air, for the men had lost the thrill of battle hours before; it was now a job, and a bloody, brutal one at that. In the distance could be heard the tat-tat-tat of machine guns and the echoing roar of the artillery pieces.

The 44th Field Artillery Battalion which came ashore immediately after the assault units was now well organized and prepared to fire barrages on any located enemy position. During the night they had moved their positions inland and continued to give fire support to the Twenty-Second.

After an unappetizing breakfast of K or D rations, the men made final preparations. Ammunition, bayonets, grenades, and rifles were checked. There could be no mistakes now; they might cost your life or the lives of your comrades.

The attack began! The Third Battalion continued to smash the beach strongpoints. Elements of the Third swung to the left to relieve units of the First Battalion around the stronghold of Crisbecq. The enemy was a stubborn and determined defender. From every conceivable defensive piece of terrain were German guns, stirring the cold early dawn with their terrifying crack. Overhead, Allied planes continued to roar inland to demolish communication systems and rear commands. It was a good sound, for with the roar came the confidence that you stood not alone in a battlefield; others were fighting too.

The Second Battalion launched their attack on the stronghold northeast of Azeville but were able to advance only a short distance. Withering enemy small arms fire was cutting the line to pieces. It was suicide to try to advance.

The First Battalion, overcoming slight resistance, moved through Ravenoville, St. Marcouf, and attacked the enemy fortification near Crisbecq. Later in the afternoon, the First and Second Battalions received counterattacks in force and were driven back almost eight hundred yards.

During the night, after the regiment had dug in, the First Battalion received a counterattack and successfully repulsed this action without a casualty. A leading factor in the repulsion of this counterattack was the accurate artillery fire placed on the advancing enemy. The artillery fired as rapidly as the gun crews could load. Firing was continuous for an hour and a half, and the ammunition expenditure was over 1,000 rounds. The gun tubes actually got so heated that the oil dripped from the guns like rainwater.

As darkness fell over the countryside, a feeling of loneliness surrounded the troops. Men prayed quietly in their foxholes, thanking God that they had survived another few hours. Overhead could be heard the artillery shells, like crumpling tissue paper, then the resounding explosion in the distance. In the mind of each individual was the common thought, "I wonder who was under that shell when it hit." In the sky could be seen the quick flash of light that the artillery pieces threw skyward when fired, then the tense moments of waiting to see if it was fired at your position. Sleep quickly overcame the exhausted men, and for a few short hours the horrors of the battlefield were forgotten.

On June 8th, the First and Second Battalions again gathered together their forces and attacked the fortified towns of Azeville and Crisbecq. Casualties mounted among officers and enlisted men, but there were no replacements arriving to take their place. As a result, hitherto unknown leaders suddenly sprang forward, and privates were commanding squads and sergeants were leading platoons. The aid men attached to the platoons worked constantly to help wounded men and to save their lives. These men with only a red-cross arm band for protection moved out under the heaviest of artillery fire to bring aid to the dying or wounded, often sacrificing their own lives in the attempt.

Company Commanders moved rapidly down their front lines, assuring the men, praising them; the officers of the Twenty-Second Infantry Regiment led their men in every respect.

The First and Second Battalions, fighting across the open terrain and taking advantage of the draws, inched forward. German coastal guns and heavy artillery fired constantly on the advancing troops. The advance was slowed, and by nightfall the objective still lay ahead.

The Third Battalion continued its mission of reducing the beach strongpoints until late afternoon. Colonel Teague was with the forward unit in order to keep abreast of the changing situation. Shortly before dark, orders were received for the Third Battalion to go into regimental reserve. Their mission was to block an expected enemy attempt to break through the First Battalion to the beaches, thereby splitting the command.

The fortress of Crisbecq had not been completed by the Germans, but it was the most formidable arrangement of pillboxes on Utah Beach. This fortress of concrete and steel, carefully camouflaged, was connected with surrounding fortresses by partially finished underground passageways. It was an excellent observation point and it was a constant source of information, by means of a deeply laid underground cable to the German defenses and command at Cherbourg. It was this fortress, surrounded by barbed wire, protected by areas thoroughly mined, that the First Battalion was to attack with unseasoned foot troops. Company A, led by Captain Tom Shields, led the attack, and as it approached the fortress, met extreme resistance immediately.

Captain Shields led the company from the front, and it was during this attack that he was mortally wounded. The Germans were meeting the attack in force and were coming toward the Company A position. Captain Shields ordered his men to withdraw, and when they tried to carry him with them in order that he might receive medical treatment, he ordered them to withdraw without him and immediately called down artillery fire on his own position, and thus lost his life in order to help save his company.

It was personal bravery of the highest order, but it was more; it was a concern for the lives of other brave men.

Task Force "Barber" under Brigadier General Henry A. Barber was formed June 9th, and the Third Battalion, less K Company, re-enforced, was relieved of its beach fortification mission and brought inland to attack Azeville. Crisbecq, still a German stronghold, was to be contained and by-passed. Company G and one company of the 899th Tank Destroyer Battalion were to do this.

In the meantime, the Fourth Engineers took over the fort at Taret du Ravenoville with orders to destroy it. The Third Battalion and its supporting units then moved back through Ravenoville and Cibrontot and attacked Azeville from the southwest. The Germans offered some resistance in the village, but in a very short time the village and forty prisoners were taken. Most of the enemy, however, withdrew into the fortifications east of the village. Company I moved on to assault the fortifications from the rear (West). Company I came in sight of the first pillbox about 1200 hours. After the artillery preparation lifted, Company I had no direct fire support except one tank—the remainder of the tank platoon did not get into action. Two assault sections, after picking their way single file through a minefield, without loss, reduced three small pillboxes on the outer ring of defenses, then the company closed in to a hedgerow about 100 yards back of the four large pillboxes. These were massive structures, with re-enforced concrete walls, six feet thick, each armed with a 150mm gun pointed toward the sea. Machine guns were mounted on top of the large pillboxes and in numerous open emplacements.

The one tank available fired 15 rounds at the nearest large pillbox, and the assault troops fired bazookas. The fire penetrated several inches to a foot in the walls but produced no useful effect. The company was under no heavy fire except three or four machine guns. Apparently, the machine guns on top of the other three fortifications were unable to fire on them. Mortar fire on the machine guns failed to knock them out. It was not until just before the strongpoint surrendered that fire from some other position to the northeast, was brought down on the assaulting troops. The supporting fires from our own 44th Field Artillery Battalion greatly aided in bringing about the seizure of the fortifications.

The attack by Company I was concentrated on the back door of the first pillbox. One assault section went in, discharged its flame thrower, and set off a pole charge against the door. It hardly scratched the door. A second pole charge also failed. Then a satchel charge of forty pounds of TNT was used. The men who set off that charge were unable to withdraw far enough and were knocked unconscious by the concussion. The door did not blow.

Captain Samuels then called for the flame thrower from the remaining assault section. Private Ralph G. Riley reported with his flame thrower. "All right, Riley," said Captain Samuels, "go in and give it a few squirts." Riley crossed the fire-swept open field for about 75 yards, reached the best position—then the flame thrower failed to ignite. Captain Samuels and those behind the hedgerows saw Riley lean over, strike a match and light the fuel. Then he emptied the flame thrower around the door and returned across the open ground. When he got back, he had bullet holes through his canteen and his gas mask, but he had escaped unhurt.

Company I had now done everything it could. Evidently the burning fuel had run under the door and set the ammunition inside afire. From behind the hedgerow, Company I could hear the small-arms ammunition popping. They waited about fifteen minutes, listening to the explosions inside. Then a white flag went up. Company I ceased firing, and the German commander came out accompanied by a non-com and an American parachute officer. The German commander, who spoke English, said that with his ammunition exploding he was ready to surrender the entire position if Samuels would cease firing. One hundred and sixty-nine men filed out. Company I in the operation had suffered only ten casualties.

Early in the morning of June 8th, Sergeant Leo of F Company was sent out on a scouting patrol with two other enlisted men. During the course of the morning, Sergeant Leo's patrol did not return. It was concluded the patrol had either been ambushed or captured, and the company commander sent out other scouts.

About two o'clock that afternoon, through the hedgerows and across the fields staggered Sergeant Leo. Two soldiers spotting him, ran to his assistance, but with a shake of his head he nodded them away and he staggered on in under his own power. The other men had been killed, he said, by enemy snipers. He himself had been wounded in the left shoulder before he could get away. His left arm, upon inspection, was very nearly severed from the shot. This had happened at about 0900. When asked where he had been since that time, he replied, "Trying to get back."

He slipped a pasteboard shell case off his right shoulder and set it on the floor while the men stared at it in amazement. It was the sort of carrier improvised and used by the medics. It was plainly marked on the top and side with a red cross and was filled with plasma, bandages, and morphine surettes. The sergeant said that on his way back, while passing a crossroads, he had seen the container, apparently forgotten, or left by the side of the road. Putting the case between his knees, he pulled the top off with his one good arm, saw that it was filled, thought the medics might need it, hung it over his shoulder, and thus had brought it all the way back to the aid station through those excruciating hours. Having told his story, having delivered his set of maps to his platoon leader, and having in a way paid for his own treatment by the medicine, Sergeant Leo went into the worst case of shock any of us could remember.

Later that afternoon, the Third Battalion advanced northwest to the western portion of Chateau de Fontenay, where again they came up on a strong enemy position. Still operating with only two rifle companies, Colonel Teague had Company I on the right and Company L on the left. Company L moved around the west of the enemy positions. I Company was stopped by the high wall of the chateau grounds which were still defended. L Company from its position was then surrounded and isolated by the Germans and was being cut to pieces. There was a wide gap between the Twenty-Second and the Twelfth Infantry Regiments, while on the right of the Third Battalion, the First and Second Battalions were further to the rear. The Germans in Fontenay Sur Mer and from other positions to the east, and from Ozeville to the north pounded L Company's positions, while enemy units to the right rear of the Twelfth Infantry were firing on them from the west. After dark, Company L was withdrawn on a line with Company I, having sustained tremendous casualties.

On June 10th, Colonel Hervey A. Tribolet was relieved of his command of the regiment. Division and Corps were dissatisfied with the progress the Twenty-Second was making. Slowed by the beach fortifications, Azeville, Ozeville, Crisbecq, and the Chateau all posed formidable barriers. The men of the Twenty-Second naturally felt they were doing their utmost. It was with genuine regret the men of the regiment learned that Colonel Tribolet had been relieved. "Trib" had trained them; he knew many of their families, and he was both loved and respected by the officers and men. Their regard for him and their confidence in his ability and leadership remained. He was succeeded in the command by Colonel Robert T. Foster.

The next three days were used in reducing the Azeville fortifications. The Third Battalion was to move toward Azeville while the Second Battalion attacked the enemy at Chateau De Fontenay, and the First Battalion attacked the town of Fontenay Sur Mer to relieve pressure on the Third Battalion, which was receiving heavy flanking fire from that vicinity.

On the morning of June 10th, the Second Battalion had been ordered to attack the Chateau de Fontenay, which had been heavily fortified by the Germans. The attack began at dawn with F Company, under the command of Captain Harold Fulton, and G Company, under the command of 1st Lt. J. O. Jackson. Captain James Burnside, commanding E Company, was the reserve company commander. Of necessity, the attack was launched across the open fields approaching the stone wall and wooded lanes leading to the chateau. German observation on our troop movements was good, and the Second Battalion soon began to take heavy casualties, although our troops continued to advance. Forward elements of Company F had reached the very walls and outbuildings of the chateau when Captain Harold Fulton and Lt. Cooke were killed and two other officers wounded. All of the men in one platoon, save only two, had been killed, and it was at this point that the order was given to the battalion to withdraw in order that the strongpoint could be bombed. The battalion was successfully withdrawn, and that night was moved out, by-passing the chateau on the east, and attacking on north to the Quinneville Ridge.

It was during the attack on the Chateau de Fontenay, once the home of Voltaire, that Lt. John Ward, forward observer for Cannon Company, was working with L Company. Lt. Ward was forward of the lines, observing, when the enemy began to come in to our position. With great presence of mind, Lt. Ward called artillery fire on his own position, in order to relieve L Company which, by this time, had been completely cut off by the enemy. This artillery barrage made it possible for Company L to establish contact once more with friendly troops, but it so severely wounded Lt. Ward that he was first taken for dead.

Captain Fulton and Lt. James Beam were approaching the Chateau and had taken cover adjoining a pump house. Captain Fulton had climbed to the top of the structure to view the enemy. Lt. Beam warned Captain Fulton to keep under cover, and Fulton dropped down with a smile and his usual ready retort just as a stream of machine gun fire was directed toward him. He suddenly stopped talking, and Lt. Beam turned toward him to find that he had been killed by a ricocheting bullet.

Sergeant Clay Edwards and Sergeant Wurtzburg were both killed as they approached the Chateau. Sergeant Edwards was attempting to go to the aid of a soldier who had been severely wounded when he was killed.

At the close of the day, the remnant of Company F was regrouped in the hedgerows before the Chateau. Only a third of the company remained under the command of Lt. Beam. Casualties had been extremely heavy and throughout the regiment, men were face to face with the fact of violent death for many of the men they had trained with for three years. It was a shattering emotional experience that hardened the attitude of the men toward such an enemy.

The following day, elements of the 90th Division took the Chateau without a shot being fired, the Germans having withdrawn. By the 12th of June, the Third Battalion had reduced the Azeville strongpoint and by nightfall had almost completed the mopping-up operations.

The remainder of the regiment was to follow up after Azeville was taken and seize the strongpoint to the northeast of there. Their attack was underway at 1800 hours, and against strong enemy defensive fires the Regiment pressed on to the outskirts of Ozeville.

Montebourg was a communication center for a large number of the enemy forces on the eastern coast of the Cherbourg peninsula. Prior to this time, several attempts had been made to seize the town by other elements of the Division, but the attacking troops had been forced to withdraw before the town fell.

Lt. Colonel Teague called together the members of his staff and carefully studied the maps, overlays, and aerial photos of Montebourg. After plans had been made, orders were issued to company commanders, and the battalion proceeded to Montebourg. They moved south of the town behind the Eighth Infantry, who held the railroad, then north along the railroad and into the town from the west. Much to the surprise of all concerned, the town had been abandoned by the Germans. Not a shot was fired. The first troops into the town believed it to be completely devoid of civilization. A house-to-house search revealed thirty Germans, seven in civilian clothes, and seven Americans wounded from Crisbecq.

Approximately three hundred French civilians came out of the cellars, having survived in the completely demolished buildings during the ten days of siege and through the terrific artillery and mortar bombardments. As soon as the town was completely searched, Lt. Colonel Teague moved his forces to the surrounding terrain to avoid an expected aerial bombing by the Germans.

Chaplain Arthur J. O'Leary, Regimental Chaplain, was relieved from duty to the regiment and returned to First Army for his new assignment. Chaplain O'Leary had served with the Twenty-Second throughout its training period, and both officers and men were reluctant to see him go.

While the Regimental CP was located near Le Theil, France, an amusing incident occurred. GIs had been thoroughly briefed on the probable attitude of the French in the Normandy Peninsula prior to coming ashore. S-2 reports indicated that French still left along the immediate beach vicinity probably would be German sympathizers and all military personnel were to be alert for attempted signaling to the enemy.

One morning a GI, with his rifle and fixed bayonet in the back of a doddering old Frenchman, marched him to the Regimental CP and on through to the IPW Team, composed of Lt. Swanick, Lt. Horne, and then M/Sgt. Sigmund Roth. The story was brief, but simple. The GI had caught the French farmer in his field attempting to signal the enemy. "How?" The IPW wanted to know. "By driving a peg into the ground and tapping out code signals," the GI replied. Upon investigation, the IPW Team discovered that the poor farmer had gone out the night before to milk Bessie, his one and only cow, only to find Bessie's milk supply gone as a result of GI's having got to Bessie first. Since discretion was the better part of valor, the Frenchman had every intention of staking Bessie out where he could keep a fatherly eye on both her and her crankcase, since not only was milk a delicacy, but steak as well. It was while he was pegging Bessie out in the orchard that the GI had caught him and promptly marched him up for questioning. The Frenchman's wife, seeing him marched off at the point of a gun, immediately concluded he was going to be shot and went into a mild case of hysterics.

The IPW Team, chuckling, sent the GI on his way and told the Frenchman to return home, which he promptly and with surprising agility did. Upon returning home, his wife greeted him with open arms, murmuring all the while loud praises to the Deity.

His return, having been celebrated with a quick slug of calvados, and with his courage thus reinforced, the Frenchman returned to his orchard and again began to stake out Bessie. While he was thus engaged, along came another GI, ran his M-1 in the Frenchman's back and marched him, protesting vehemently with arms and words, back to the IPW Team. The incident this time was not so amusing to the IPW Team, since the story was the same, and besides, they had many German prisoners to interrogate, so they sent a guard along with the old man with instructions to personally supervise the staking out of Bessie and to assure the completion without further interruption. This the GI did and happily enough since he was an honored guest at the resulting second calvados celebration.

After the fall of Quineville Ridge and Montebourg, the

Twenty-Second moved steadily toward the northwest corner of the Contentin peninsula. Casualties were much less, and resistance was crumbling.

Captain Howard Blazzard was having a great time firing German 88's back at the enemy. The Germans often removed the firing pin from an 88, taking it with them. In the event they could recover the position they lost, the pin could be replaced quickly, and the gun used again.

Captain Blazzard found a firing pin intact, and carried it with him, much to the amusement of some of the other officers. But approaching the airport near St. Pierre Eglise, he found excellent opportunity to use it. Popping the pin in place in a particularly well-placed gun, he had a wonderful time using German ammunition and a German gun to shoot Germans!

Casualties were heavy approaching the airport as a result of ack-ack fired at short range. Cherbourg was the objective of the Fourth Division, but the Twenty-Second moved to the right of the Twelfth Regiment, securing the eastern section of the peninsula.

Cherbourg fell on 25 June; the harbor was in poor shape and must surely have been a sharp disappointment to the Allied Command since it was to be useless for shipping for many weeks. The Twenty-Second continued to fight for two additional days toward St. Pierre Eglise. As a result of first-rate intelligence work and coordinated inter-unit information, the rest of the enemy resistance on the peninsula crumbled, resulting in the surrender of more than 900 German officers and men. The conquest of the Cotentin Peninsula was complete, just twenty-one days after D Day.

The fighting had been bitter and costly. Three hundred and seventy-three enlisted men had been killed and fifteen hundred and sixty wounded. Twenty-three officers had been killed and one hundred and four wounded, for a total of two thousand and sixty casualties in less than one month of fighting, in a regiment whose normal strength was about three thousand two hundred men.

6. Periers — Corridor Of Death

In spite of the irritation of training, the relaxation was wonderful! It was a good feeling to sleep without being shot at! The first hot meals since England were served, and never had food tasted so good! Hot showers and the first change of clothing did much to raise morale. The first mail since the invasion was delivered, and the joy of word from home was dulled by the number of letters that had to be returned marked WIA or KIA (wounded in action or killed in action).

On July 7th, the regiment prepared for an attack on the town of Periers. Assembled at Amfreville, the command was warned that fighting would be bitter and opposition severe, since the main German line partially blocked the American forces in Normandy. Feeling was strong the enemy would attempt a breakthrough to the sea, thus cutting the Americans off from their base of supply.

The fighting began on the morning of July 8th and soon became one of the bitterest battles of the war. French farmland was divided into sections by hedgerows, grown up many feet and providing excellent cover and concealment for the enemy.

The mission of the Second Battalion was the breaching of the enemy line south of Culot with La Maugerie as their objective. Neuville was the objective of the First Battalion.

The Second Battalion, after passing through lines of the 83rd Division in the morning, had been advancing against stubborn German delaying action all day. In the late afternoon, they had been stopped by strong resistance about 500 yards southwest of the creek at Culot. About 2100 hours, they were ordered to secure lines for the night.

About 2130, three German tanks accompanied by infantry came up the road against the Second Battalion's position. The leading tank continued up the road between Companies F and G toward the rear of the battalion. At the road junction a few hundred yards in the rear stood an American tank which had been knocked out and was blocking the road. Lt. Colonel Wellburn, CO 70th Tank Bn had just come up to examine this tank when he heard the German tank approaching. At the same moment, Lt. Colonel Wellburn saw an American half-track towing a 57mm gun coming down the road behind him. He stopped the half-track and said, "Uncouple that gun. Here comes a Kraut tank." The driver immediately pulled into the farmyard just behind the wrecked tank and the gun in a few seconds was uncoupled at the corner of the gate, covering the road.

As the German tank came around the bend in the road, he saw the American tank a bare hundred yards in front of him. The German stopped and opened fire. His first round set the American tank on fire, and he moved forward slightly and continued firing. Several rounds went through both sides of the turret of the American tank, and one went all the way through the tank and knocked the door off the back.

Meanwhile, the 57mm antitank gun was loaded and ready to fire. As the side of the tank came into view, the sergeant said, "Pour it in." The first shot hit the most vulnerable point and knocked the German tank out. The 57 put five more shots in the same place. When this encounter ended, the two tanks, German and American, were facing each other, 50 yards apart, both riddled with holes and their crews wounded or dead.

A second German tank accompanied by infantry had swung west into the orchard in front of Company F, opening fire on the company. At the same time, the Germans put down a heavy artillery barrage on the orchard.

Another German tank had swung east into the orchard in front of Company G. Private Hicks, with a bazooka, stood at the corner of a small house near the left flank of Company G and fired at the approaching tank. He got four hits, and on the fourth the German tank blew up. The turret was blown off, and the tank tipped over.

With two of the German tanks knocked out, the third withdrew. Company F was immediately ordered to retake the field from which they had withdrawn, and Captain Tommy Harrison, Battalion S-3, moved up to tie in the lines and prevent gaps from forming. The 44th Field Artillery, in direct support, laid down such a heavy barrage on this field and the enemy positions behind it that the smoke completely obscured the scene. Company F attacked just before dark but was stopped by heavy German fire. About 0230 the next morning, Lt. Clark and six men who had been in the disputed field throughout our bombardment, succeeded in returning to our lines.

Meanwhile, the First Battalion moved through the gap effected in the enemy lines by the Second Battalion and before dark was on its objective.

It was well after midnight before all the troops were situated and dug in. Sleep was difficult. It was rumored throughout the command that the Germans had forty tanks in Periers which they intended to commit some place on the line. Tank traps were constructed in every conceivable approach and these traps covered with bazooka teams. The battle had only begun, but the replacements knew what to expect. Darkness on the front was always a time of discomfort and tension. Water seeped into the foxholes and clothing was clammy.

With the dawn, radios and field phones hummed with pre-attack conversations and last-minute orders. The regiment was facing continued stiff resistance. Once again, Division HQ was unhappy with the failure of the regiment to attain its objectives and to break the German resistance. Accordingly, Colonel Robert Foster was relieved of his command, and quite unknowingly, the Double Deucers were about to begin a shot-gun association with a colorful figure who was to change the destiny of a regiment.

A field phone rang. A Captain answered and heard, "I am Colonel Charles T. Lanham. I have just assumed command of this regiment, and I want you to know that if you ever yield one foot of ground without my direct order, I will court-martial you."

It was a proper introduction to "Buck" Lanham. But the regiment began to fight with a skill, imagination, and daring it would not have attempted before.

The 83rd Infantry Division was on the left flank. The usual rivalry between units became apparent when the Second Battalion of the Twenty-Second, moving into the attack, found their left flank exposed through the failure of elements of the 83rd to move forward simultaneously with them.

"Continue the attack" was the order on the 10th with the First and Second Battalions in the assault. As the attack moved ahead, the Germans fell back to successive hedgerows, intending to fight a delaying action until they could get their lines organized well enough to smash the oncoming assaults.

About midday, the Third Battalion was committed on the east (left) of the Second Battalion, with orders to seize that portion of the Regimental objective east of La Maugerie. Shortly before dark, the Third Battalion's movement across the front of the Second Battalion masked the fires of that unit, and the Second Battalion reverted to regimental reserve. The First Battalion advanced to the outskirts of La Maugerie and held that position for the night, with the Third Battalion on their left.

It was during the bloody La Maugerie battle that the mortar barrage fell in the Company F area, wounding Lt. Jim Beam, the company commander, when a mortar shell landed almost between his feet. This was the same Jim Beam who, a couple of days earlier, upon receiving a radio message from Major Edwards as to his situation, replied, "The enemy has broken through on our left flank. They are infiltrating and attacking with tanks on our front; our right flank is exposed. The situation is well in hand." Now with a right foot almost severed and with a left foot badly broken, Lt. Beam calmly lighted a cigarette and assisted Captain Humm, the Bn. Medical Officer, in completing the amputation of his foot. Having given himself a morphine surette, he refused to be evacuated until he had completed his orders, reorganizing his company to withstand the fierce German attack. Such were the men of the Twenty-Second Infantry.

Again on this day, Private Hicks, who was then known as the "human tank-destroyer," got his third tank from the corner of a hedgerow behind a tree. He got three bazooka shots into the German tank at less than five yards. He was so close that the explosion scorched his face.

Lt. J. O. Jackson, tired of the constant harassing and dangerous fire from an enemy machine gun, climbed out of his foxhole and, before the eyes of his astonished G Company, crawled quietly around the edge of a field, over a hedgerow, pulled a pin from a hand grenade with his teeth, threw the grenade at the machine gun nest, and immediately after its explosion, rushed in to polish off the Germans with his bayonet, and then climbed back calmly over the hedgerow and back into his foxhole as he remarked succinctly, "That's the way to do it."

On the 11th of July, the attack was resumed. The Third Battalion was directed to seize the high ground in the vicinity of Raids by envelopment from the east. This battalion struck a strong defensive position and was delayed. The Second was employed to their left, and Company C mopped up La Maugerie to the southeast.

The attack on the 12th was to prove extremely costly in lives of both the officers and enlisted men. The Germans were waiting and well prepared.

The first attack gained about three hundred yards following a heavy artillery preparation. Forward observers from the 44th Field Artillery Battalion moved well out in front of the lines in order to obtain good observation posts to better direct the fire of their units. It was an example of bravery doubly appreciated by soldiers who knew the effectiveness of artillery fire. The attack had only begun when Captain James B. Burnside, Second Battalion Executive Officer, was wounded, leaving the Second Battalion with only seventeen officers.

The fighting became fierce hand-to-hand conflict. Casualties mounted rapidly; the rifle companies did not have over seventy effective fighting men. Though the number was small, individual courage and initiative were everywhere apparent. First Sergeant Kenyon, of G Company, gathered together fifteen men and took over a section of the front normally held by a platoon. He said in a rather calm undisturbed manner, "I've taken over this part of the front and I'm going to hold it. You don't need to worry about it."

The Battalion Commander, Lt. Colonel Earl W. Edwards, was inspecting his battalion following one of the fiercest bits of fighting on the 10th of July, reorganizing and tying in. While crossing a field from one hedgerow to another he came upon a medic, a private first class, who had been on the front lines every single hour since the battalion had landed. His Company Commander swore by him, and everybody talked of his courage. But he had had about as much as he could take. When the Lt. Colonel approached, he was doing his best to aid one of those impossible tragedies of the war, a man whose skull had been laid open and who had sustained such a severe brain injury that he could not possibly live, and yet he would not die. The medic had done everything he knew how to do. He had been alone on this field for so long and the strain had been so great, and yet a devotion to duty and to a wounded member of his company was so great that he would not leave.

When he saw the colonel, his reserve broke and he cried, "Sir, he won't die, He ought to die, but he won't. I have done everything I can for him, but he won't die. Why won't he die?" The colonel led the aid man away, giving instructions for much needed rest and care for the boy. This was war at its lowest, purest hell.

By the morning of 13 July, the corridor of death had taken its toll. Little was left of a proud fighting regiment that had faced a determined enemy. The battle had been almost as costly as the beachhead assault. One thousand forty-four enlisted men were wounded and 263 were killed. Fifty-six officers were wounded and 16 were killed, for a total casualty list of 1,379 in less than one week of fighting. Total casualties for the five weeks now stood at 3,439, more than the total strength of the regiment at D Day.

In the Battle of Periers, even though the regiment did not succeed in capturing the city, they did help to break the back of the German defense of this sector in the bitterest hand-to-hand combat of the war.

During this period, Chaplain William S. Boice, Corporal Otto Oehring, and American Red Cross Field Director David Mitchell drove to Cherbourg to obtain flowers and wreaths. These flowers were placed on the graves of all known Twenty-Second men in the St. Mere Eglise cemetery. It was a sad task, done out of regard and genuine affection for fine men of a great regiment! But the crosses and Stars of David were many and presaged many more to come in the bitter fighting ahead.

Some period of rest and regrouping was essential, and while only in reserve, the command set about rebuilding the regiment into a crack fighting force. That this was quickly and effectively done became at once apparent, for a Combat Team was formed for a special purpose. Consisting of the Twenty-Second Infantry, the 44th Field Artillery, Company C of the Fourth Medical Battalion, the team was formed on 19 July, moving east to the vicinity of La Mine. Here it became a part of Combat Command A, 2nd Armored Division, under Brigadier General Maurice Rose.

The Allies were gathering their forces in the peninsula. General Patton was poised for a thrust into Normandy. Tension mounted as the regiment sensed they had been given a special assignment of vital importance. What that action was to be lay shrouded in the clouds that prevented saturation bombing that was to presage the attack. The regiment was ready!

Six officers and enlisted men of the 22nd Infantry being awarded
the Silver Star by Major General Raymond O. Barton in France
July 17, 1944 — U.S. Army Signal Corps photo SC-244296

7. Vive La France — On To Paris

"Sharkey made the greatest display of guts I've ever seen."
— Captain Frank Reid

"That artillery fire was beautiful."
— Lt. Gerald J. Claing

By 25 July, the beachhead of the Allied Armies in Normandy had been secured, and enough men and material landed to begin a large-scale offensive to liberate France.

The military plan was to shift from the hedgerow, hand-to-hand fighting of the beachhead, to a new phase which would use the full and coordinated power of air, artillery, infantry, and armor.

The enemy had drawn its forces on a line from Coutances to St. Lo. To affect a break-through of this line was essential and it was to this task the Second Armored Division with the Twenty-Second Infantry providing the infantry support was sent.

The men had been rapidly and well trained in the intervening days in coordinated fighting. Each platoon of infantry drilled with the platoon of tanks with which it was to fight, learning how to protect the tank while using the fire power and cover of the tank as supporting fire power for their advance. The performance of the infantry during the battle for

St. Lo proved the effectiveness of this training and concept of fighting.

The Allied Command had ordered saturation bombing of the area to be attacked prior to the attempted breakthrough. On 25 July 1944, the weather cleared, and the bombing began. It was deep enough and wide enough to shatter enemy resistance to the armored spear we were to send through their lines. Double Deucers counted as many as sixteen bomb craters per acre. Houses swayed as if made of some macabre papier-mache. Mark V tanks were turned on their backs like toys, with the occupants killed or stunned by the concussion.

German soldiers walked dazedly, unresisting, and unknowing, their eyes glassy and mouths gaping. More planes flew overhead, and enemy soldiers tried to hide from the death from the skies, but in vain!

Combat Command A was waiting behind the 30th Infantry Division to go through the gap as soon as it was made and exploit the break-through. No advance was made by the 30th Division on that day, however, and the break-through had not been completed. Therefore, Combat Command (CC) A was ordered to pass through the 30th Division on 26 July to complete the break-through in conjunction with the balance of the Fourth Division and First Division and then to continue the exploitation.

Fifty-three minutes past midnight on the morning of 26 July, the regiment, as part of CC A, began leaving its rest area in the vicinity of La Mine to begin its break-through operation in the direction of St. Gilles and Canisy. Originally, the CC was to attack in two columns, with the First Battalion in the left column and the Second and Third Battalions in the right or north column. The Third Battalion, in reserve behind the First, was to move by bounds with the special units. However, during the night of the 25th, plans were revised by VII Corps, and the attacking force was changed to a single column with all elements on the north route.

The CC moved to a forward assembly area two miles north of Pont Hebert and attacked to the south at 0930 hours, with the First Battalion, Twenty-Second Infantry, on the tanks of the Second Battalion, 66th Armor. Initially, the movement was via hedge rowed fields and followed the following tactics: A tank platoon and a platoon of infantry would work together as a team. The five tanks deployed, with two going ahead as scouts and three following as a covering force. Upon entering a field, the two forward tanks would spray the hedgerow on their open flank and then would fire a 75 mm round into the far corner of the field. At the same time that they crossed the field, they fired their machine guns at the hedge to the front. The covering force tanks followed and drew up online with the others and the doughboys trotted past to reconnoiter the hedge in front of the tanks for German bazooka-men and for the best places to break through the hedge.

After this reconnaissance, the doughboys ran back behind the tanks and the two "scout" tanks would start the process again. The advance was slow at first, due to unexpectedly heavy resistance from enemy positions which were still holding out. The Command encountered considerable high velocity artillery shelling but very little, small arms fire. Nevertheless, the effectiveness of the great aerial bombardment on the preceding day was clear. The Germans were still shaky, and the prisoners seen going to the rear looked beaten and stunned.

As the day progressed, the advance was accelerated, the CC having cleared the bombed area and apparently most of the German resistance as well. Late in the afternoon, CC A was directed to revert to the two-column plan and to continue its advance on roads in order to accelerate the advance. Owing to the great length of the column, the disposition then in effect, and badly cratered roads, this split was not completed until St. Gilles was secured.

It was approximately 2100 hours when the CC passed through St. Gilles. Normally it is considered that tanks cannot operate successfully at night, but General Rose, Commanding CC A, was determined to take the objective regardless of time and the cost in men and tanks, since the whole operation depended upon the success of that mission. The advance continued down the main road to Canisy in complete darkness. As the CC passed through Canisy, the whole west side of the town was burning fiercely, and the long tank columns going through silhouetted against the blaze made a memorable scene. The Air Corps had functioned superbly throughout this advance, working just ahead of the CC with perfect coordination all the way. They were very careful to be sure of identification before they bombed vehicles on the road. On one occasion, dive bombers came down twice for a close look at enemy vehicles and then the squadron leader himself dove down to be positive that they were German before attacking them.

A break-through had been affected despite German infantry and armor. St. Gilles and Canisy had been taken! During the remainder of the night, the head of the column occasionally stopped for some resistance, but it was never held up long. There is an amusing incident connected with this stage of the advance. Just north of Canisy, a German half-track with a trailer pulled into the column near the head and passed several tanks without being detected before making the mistake of passing the leading tank. The furious tank commander exclaimed, "Hell, I'm supposed to be leading this column," whereupon he recognized the vehicle as enemy and opened fire, knocking it out.

By dawn of 27 July, CC A was on its objective, the key terrain in the vicinity of Le Mesnil Herman. These positions, including Hill 183, Le Mesnil Herman, and St. Samson De Bon Fosse, had been secured by 1800 hours on the second day. Approximately 300 prisoners were taken, and the advance had gained 10 1/2 miles. As the regiment dug in for all-around defense, reconnaissance teams were pushed to the south. Infantry patrols combed the area. Late that afternoon two task forces, each consisting of a company of tanks and a company of infantry, were dispatched to reconnoiter in force to the south. Company K, in the western force, drove as far south as Villebaudon, where it was picked up the next day by the remainder of the Third Battalion attacking south to Percy.

During the morning of 28 July, the regiment consolidated its position and maintained enemy contact with patrols. That afternoon the CC issued orders for a three-column attack. In compliance with these orders, the Third Battalion was attached to the Third Battalion of the 66th Armored Regiment and pushed south along the Le Mesnil Herman-Percy Road. The First Battalion with the attached armor, attacked south in the direction of Moyen; the Second Battalion on their left (east) moved toward Tessy Sur Vire. The CC encountered a regimental combat command of the 30th Division delayed by a determined enemy. One effort to break the position was thwarted by a large stream, and the task force dug in. At the close of the day, CC A had halted with the Third Battalion near Percy, the Second Battalion a mile and a half south of Le Mesnil Herman, and the First Battalion occupying Moyen.

Two men of the First Battalion, Captain Frank B. Reid, CO, Company C, and Pvt. Sharkey, also of Company C, merit special attention for their heroic actions during the Battalion's attack on

Moyen on 28 July. Undoubtedly there were many other similar performances deserving of equal praise, but the story of Captain Reid and Pvt. Sharkey is indicative of all the fighting men who distinguished themselves so remarkably in the face of almost insurmountable obstacles.

After penetrating the northern part of the city of Moyen on the 28th, the Battalion discovered that the enemy had strong tank forces south of the town along with other entrenchments and antitank guns. Captain Reid led a patrol of twelve men south to a crossroads just east of the church to get a German antitank gun. The crew of the gun was dispersed by rifle fire and grenades, but immediately the patrol received heavy machine gun fire from a German tank in the distance. Captain Reid moved his patrol by circuit to the north and east and came up on the right flank of the tank. The ammunition carriers for the bazooka had been lost by this time and the patrol had only two rounds of bazooka ammunition.

Pvt. Sharkey, known as a "bazooka hound," fired at the tank just across the hedgerow and hitting it in the turret, knocked it out. A second enemy tank now came up directly behind the one destroyed and started blazing away from a distance of from 75 to 100 yards, whereupon Captain Reid leaned over the hedgerow just above the tank and dropped two WP grenades, one down the air vent on top of the tank and the other under the tank, setting it afire.

A column of German tanks, accompanied by infantry, was now coming along the road towards the town, and they opened a terrific volume of fire on the patrol. Pvt. Sharkey stood on top of the hedgerow and fired the last bazooka round at the leading tank, hitting it at the base of the turret and knocking it out. "Let's clear out of here before they zero in on us," Captain Reid shouted to his men. But Pvt. Sharkey remained standing on top of the hedgerow and fired at the approaching infantrymen with his carbine until a burst of machine gun fire from the tanks took off the whole side of his face. It missed the bones of the jaw but left the flesh hanging down over his chest. Pvt. Sharkey got up from the ground and walked away with the retreating patrol. The rest were crawling across the field under the heavy fire from the German tank column while Sharkey made the whole trip walking upright.

As they reached the road running east from Moyen, they found that another column of German tanks had moved up that road toward the town, and the leading tank was barring their way across the road. The patrol now had no ammunition other than two WP grenades, one fragmentation grenade, and small arms ammunition. Captain Reid threw the two WP grenades, one under the tank and one behind it. The tank backed up, but it was already on fire. Meanwhile, the smoke from the grenades formed a screen across the road and under this cover the patrol got across and rejoined the company, without Sharkey and two other men whom Captain Reid had left to take care of Sharkey while he got the remainder of his patrol to safety.

By this time, Sharkey had practically collapsed, and in the brief duration of the smoke screen, the other men were unable to lift him over the hedgerow. However, the German tanks had apparently been successfully frightened away; a little later the three men moved down the road into town, Sharkey again walking and holding up his fingers in the victory sign. "Sharkey made the greatest display of guts I've ever seen," was Captain Reid's comment on the incident.

All units advanced at daylight on the 29th. The left and center columns renewed their attacks but without success. At Moyen, the tank commander of the Second Battalion, 66th

Armored, attempted to plunge through the town, but German fire knocked out one tank on the road to the southeast and another on the road to the south, thus blocking both roads. The Germans had tanks dug in south of the town as well as a strong force of mobile tanks; a prisoner subsequently stated that they had eight Panzer companies in that sector. When the armored attack had failed, the tank commander decided to withdraw from the town and shell it. The withdrawal of the infantry was already under way when Major Latimer, the First Battalion Commander, discovered it and persuaded the tank commander to countermand the withdrawal, the tank commander failing to understand the harmful effects of infantry withdrawing from ground once won. The battalion tried to move back into the town, but the Germans had followed closely behind them as they withdrew, and they never succeeded in regaining all the ground given up.

The Germans now brought up strong tank forces just south of the town in addition to those that were dug in there, and a duel developed between the opposing tank forces, with the infantry in between. It was a terrible experience, and losses ran very high. Our forces were also under a great deal of artillery fire. In addition to the heavy physical casualties, both infantry and armor had a number of men who cracked up under the strain. One German tank which came up on the southwest caught the right flank platoon of Company A under very heavy fire and wiped out all but eleven of its men.

The Second Battalion, which was teamed up with the First Battalion of the 66th Armored, encountered strong resistance around Bessinerie and was stopped in a position abreast of the First Battalion on the road leading south to Tessy Sur Vire. When the Third Battalion fought its way to the high ground

1200 yards north of Percy, it was seven miles to the south of the remainder of the CC, and thus partially isolated.

At 1700 hours, the three battalions of the Twenty-Second were attached to the armored battalions with which they were fighting and so passed from direct control of the regiment. Later in the day, this task force was relieved at Moyen by the 116th Infantry, and they withdrew just in time to escape a bombing by the Luftwaffe in which the Germans hit their own lines as well as ours. The CC might have gone unobserved had not a machine gunner in one company cut loose with his gun at the planes. This was unfortunate, for the planes began circling and dropping flares which lighted up the landscape "bright as day." These flares were followed by bombs, but most of the "heavies" fell on the Germans who were a few fields away, due probably to the fact that the flares had drifted back over the German lines, thus causing the bombing error. The German infantry began frantically shooting up green flares to stop the bombing. The incident pleased the troops of the Command a great deal, and as Major Latimer, First Battalion Commander, later remarked, "We were glad to see that the Germans could make mistakes too."

Because of this bombing and the shelling, which was still going on at frequent intervals, the relief was not entirely completed until after midnight. At that time an attempt was made to move out on the road running west from Moyen, but the Germans were also intrenched there with dug-in tanks, and this move failed. The entire task force withdrew to the northwest and bivouacked for the night a mile or so south of Le Mesnil Herman.

Meanwhile, the right task force, including the Third Battalion, had resumed its advance at 0700 that morning, Company L riding the tanks, and Company I following on foot. Company K was still holding east of Villebaudon. Company L got as far as La

Tilandiere where they were stopped by a strong enemy position along the sunken road. Company K had been relieved at Ville-baudon by a battalion of the 175th Infantry and had rejoined the Battalion. Later in the day, Company I was relieved by two companies of the 29th Division, while Company L continued to contain the enemy at La Tilandiere until the next day. The rest of the task force withdrew to Mesnil Ceron, where they bivouacked for the night.

The day ended with elements of CC A and the 29th and 30th Divisions in contact with strong enemy positions on a discontinuous line from Bessinerie to Percy. Two days of fighting had failed to dislodge the enemy from any of these positions.

On 30 July, the First Battalion was committed in the vicinity of Le Denisiere, where enemy forces had been cutting the supply lines of the Third Battalion. Along with the Second Battalion, 66th Armored, the Battalion moved south on the secondary road paralleling the highway on the west, which was under heavy observed artillery fire from the German positions to the east. The force reached Villebaudon about noon, just as a strong German tank force, possibly the same Panzer unit which had been met the previous day at Moyen, was overrunning the town. Under the pressure of the deployed American tanks and tank destroyers, the Germans withdrew on the north. At the same time, Company B, Twenty-Second, and a tank company on the right flank drove the enemy back in the south, while still farther south, other tanks had a considerable fight with Panzers which were moving westward. The task force suffered heavy tank losses in this battle, but the German threat was broken.

The Second Battalion, Twenty-Second, with the First Battalion, 66th Armored, having been relieved at Bessinerie by the 116th Infantry, had moved to an assembly area near Le Mesnil

Herman, where, on the night of 30 July, it was subjected to bombing from the Luftwaffe.

The medical section was hardest hit in this attack. One bomb, landing close to a hedgerow, blew three medics into bits. Ironically, these men, of necessity, had to be listed as "missing in action" since there was literally nothing at all left by which they could be identified.

The ammunition dump was also hit, and a half-track loaded with ammunition set on fire. A colored private, later recommended for the Silver Star, went into the burning ammunition vehicle to drag out a Quartermaster officer who was lying under the half-track but who, as it turned out, was already dead. The total casualties in this attack were six dead, twenty-five wounded, and seven vehicles lost.

Meanwhile, the Third Battalion was in contact with enemy forces north of Percy, but the situation there did not materially change. The Battalion received a terrific mortar barrage, causing many casualties, and in addition, lost seven tanks, probably by bazookas.

Thus, at the end of the day, CC A had one battalion task force overlooking Percy, one defending Villebaudon, and one in reserve. Various units of the 29th Division were interspersed in the same area.

On 31 July, the armored columns, of which the First and Second Battalions were a part, engaged the enemy around Villebaudon, while the Third Battalion still fought on the northern outskirts of Percy. Late in the day, the armored combat team, which included the Third Battalion, was relieved of its positions and returned north to rejoin the balance of CC A. Thus, the Command was together again in the Villebaudon area. The entire 17th Infantry was also assembled there during the day.

During the night of 31 July—1 August, the Combat Command was attached to the 29th Infantry Division, and orders came down to attack and seize Tessy Sur Vire the following morning. The plans called for the Second Battalion and the Third Battalion to attack abreast, with the Second on the right. The First Battalion was to be in reserve. A severe infantry-tank battle followed. The Germans put up a fight in front of the First Battalion and several times deployed tanks across the road to stop them. This was combated by the well-tried method of infantry dismounting, locating the enemy tanks, and then bringing up our tanks and TD's (tank destroyers). Seven German tanks and an antitank gun were knocked out by the First Battalion task force in this manner. By late afternoon the Third Battalion, with its accompanying armor, had taken Tessy Sur Vire and had outposted the surrounding high ground.

During the forenoon of August 2, the 30th Division relieved the rifle battalions of the 22nd Infantry near Tessy Sur Vire and the Regiment reassembled at Villebaudon. At noon of that day, the Regiment was relieved from attachment to CC A and returned to the control of the Fourth Division. The Regiment, immediately upon relief, moved by truck convoy southwest to the Fourth Division sector via Villebaudon, Hambye, Villedieu, to Les Poelies, where the First Battalion with Cannon Company and Antitank Company had, prior to their arrival, established a defense of the town. The remainder of the Regiment, in Division reserve, bivouacked just north of the town, closing in this area early in the morning of the 3rd of August.

At 1300 hours the Regiment moved south, the First and Third Battalions abreast, to mop up the division zone of action in the wake of the 8th and 12th Infantry Regiments. The advance moved rapidly against negligible resistance. At 1700 hours, the movement was halted. The First Battalion shifted to La Chapelle Cecelin and constructed roadblocks to stop any attack from the coast or the northeast. Antitank Company and one platoon of Company C, 634th Tank Destroyer Battalion, assisted in the mission. The Second Battalion moved to the high ground north of St. Pois and outposted to the southeast. The Third Battalion, in reserve, moved to the vicinity of La Gurie.

Positions remained stationary with no enemy contact until 1300 hours 4 August, when the First and Third Battalions moved to forward assembly areas 3,500 yards northwest of St. Pois. The Second Battalion remained in place and reverted to regimental reserve. The First and Third Battalions attacked in line, the First on the right, to seize the town of St. Pois and the high ground to the east and southeast of that town. The attack met a shower of small arms and tank fire from Hill 232 north of St. Pois, thereby preventing an enveloping movement from the left. By dark, the First and Third Battalions were generally along a line 700 yards northwest of St. Pois.

During the night 4-5 August, Colonel C. T. Lanham, Regimental Commander, requested and received authority to attack Hill 232 in the 12th Infantry zone with the Third Battalion. This attack got underway at 1025 hours, and by 1900 hours the objective was taken. The First Battalion followed up, attacking to the southeast, by-passed St. Pois, and by 2000 hours had secured the woods and high ground east and southeast of St, Pois. Patrols from the Second Battalion, at that time in reserve, entered St. Pois, and by 1800, Company E re-enforced, occupied the town.

The First and Second Battalions patrolled the regimental zone of action during the night of 5-6 August with negative results. At 1000 hours the Second Battalion attacked in approach march formation, meeting only slight resistance. By 1300 hours the en-

tire assigned area, from Hill 232 as far south as Chateau Lingeard, was cleared of enemy troops. Units of the Twenty-Second
then moved to an assigned rest area north of Cuves. A composite
battalion, under command of Major Glenn Walker, Commanding Officer of the Second Battalion, consisting of Companies E,
F, and K, re-enforced by heavy machine guns and one platoon of
antitank guns, remained in positions to outpost the high ground
from Hill 232 southeast to Chateau Lingeard.

In this connection, special tribute should be paid to the 44th
Field Artillery and to the mortars of Company F for outstanding performances in the action around Chateau Lingeard on 6
August. The 44th F. A. from positions west of St. Pois was prepared to fire concentrations on Lingeard, La Cheminee, and on
a stream junction east of Hill 230. No rounds were fired to sight
the howitzers in, but all fires were plotted by coordinates. Later,
when another company entered Lingeard, they found 90 dead
Germans and captured 60. A number of these men had been
killed by perfect artillery fire on the prepared concentration. At
Lingeard, the Germans had had an assembly area, perhaps for a
battalion. "That artillery fire was beautiful," remarked Lt. Gerald
J. Claing. The first rounds were perfect hits, despite the fact that
the concentrations had only been plotted by coordinates.

As for the mortars, it seems that at a meeting the day before,
the battalion commander had made a strong recommendation to
his men to use their mortars more, since everyone had noted the
neglect in using the 60's since D-Day. Apparently, the advice was
taken seriously, for on the following day the mortars of Company
F were used extensively and to great advantage. During the advance on the morning of the 6th, movement was held up on two
occasions by two enemy machine guns, both of which were eventually knocked out by 60mm mortars. About noon, when the

company was setting up defensive positions, another machine gun opened up on the platoon which was on the lower slope of the hill. The same mortarman who already had two machine guns to his credit came to the fore once more to make his total three.

About 0200 hours, the Germans threw their first counterattack, and coming from Lingeard, they moved right up into the creek bed below the hill. One of the outposts reported back to the platoon leader who passed the word on to the weapons platoon leader. The company order had been that if a counterattack should come from the little valley, all small arms fire would be withheld while the mortar would knock out the attack. The weapons platoon leader then got Sgt. John Prettyman, mortar squad leader, to lay down a barrage on the creek bed. There were 70 rounds of ammunition on the position. Lt. Claing, who had been called at his CP back off the road, ordered the jeep to carry up its load of an additional 70 rounds.

Ammunition was going so fast that it was finally necessary to send back to battalion supply at St. Pois for more. That night the Germans probed against F Company's position, approximately one patrol each hour. Everyone was repulsed by this one mortar alone which fired a total of 370 rounds with only two misfires. All night long, Sgt. Prettyman rained 60mm mortar hell down on the Germans. The German attack began with a 50mm mortar barrage which crept horribly right around the little road on top of Company F's hill. They attacked with machine guns and "burp" guns but could do nothing against the one mortar. The Americans fired not one single round of small arms fire that whole night and suffered only one casualty (one man was shot accidentally earlier in the evening). The mortars had proved themselves!

Before dawn the 7th of August, Company 'F', deployed on the high ground in the vicinity of Chateau Lingeard, repulsed two enemy counterattacks. Company 'C' re-enforced with one platoon of heavy machine guns, one platoon of antitank guns, one company of the 70th Tank Battalion, and one platoon of tank destroyers, was alerted to seize and hold Hill 230 north of Chateau Lingeard, the objective of the 47th Infantry of the 9th Division. The attack moved out late at daylight, but the objective was secured by 1045 hours.

During the early morning hours, an enemy attack broke through the 39th Infantry area and penetrated through to the area held by the 22nd Infantry. Company 'B' and Antitank Company, less two platoons, moved south of Cuves to construct road-blocks on all roads and trails from the south and east. Company 'A' was moved from the rest area to a position south of Chateau Lingeard to add greater strength to the outpost line in that sector. Company 'K' was relieved by elements of the 47th Infantry and moved from Hill 232 to the rest area, reverting to Third Battalion control.

On August 8th, separated elements of the 9th Division joined southeast of St. Pois and pinched out the Twenty-Second Infantry. To meet this situation, the Regiment re-aligned in a defensive position along the line Chateau Lingeard—St. Pois backing up the 9th Division. The units comprising the composite battalions were relieved and joined their battalions. There was no direct contact with the enemy that day or the next, and the Regiment remained stabilized. But this turned out to be the calm before a storm, and about 2100 hours the Regimental CP was sprayed with artillery and mortar barrages which wounded the Regimental Commander, the Regimental Executive Officer, the Headquarters Commandant, killed Mr. Harvey, the Assistant Adjutant, and killed or wounded some fifteen additional men in the area. The First Battalion was placed on an alert status at 2255 hours for possible movement to aid the 30th Division.

Again, on August 10th the situation was quiet in the Regimental area except for some artillery fire. At 1500 hours an order was issued for early movement south to the vicinity of Le Teilleu on a defensive mission. The Regiment started moving at 1730 via St. Pois, Brecey, St. Hilaire de Harcout, Buais, to Le Teilleu, with the leading company on tanks and the balance by motor convoy. The thirty-five-mile movement, completed at 2330, was rapid and without enemy contact. To thwart any possible enemy thrust, a defensive position was established from Passais north along the west bank of the La Varenne River to the vicinity of La Bourdonierre.

At daylight on August 11th, vigorous patrols moved east toward the La Varenne River to ascertain the enemy's disposition. Patrolling continued throughout the day and confirmed reports that the enemy positions were along the east bank of that river. Enemy patrols were encountered on the west bank of the river, and some prisoners were taken. The day was used to improve defensive positions, with particular emphasis on blocking roads that might be used by enemy armored columns.

The main defensive positions were quiet on the 12th, but active reconnaissance patrols made several contacts with the enemy forces on the La Varenne River. Regimental patrols reconnoitered to a depth of 5,000 yards inside the German lines, despite considerable numbers of the enemy armed with rifles and machine guns who were laying for the patrols. Later in the day the defensive sector was increased, and Company 'E', re-enforced with one platoon of heavy machine guns, and Antitank Company, less one platoon, was shifted to Barenton with the mission of establishing a defense around the town.

Possibility of an enemy offensive seemed remote by the 13th, and there were definite signs of enemy retrograde. That afternoon a combat patrol from the First Battalion established a small bridgehead across the La Varenne River and, after a light skirmish, occupied Torchamp. That same afternoon, friendly troops passed through the lines moving east and this, in conjunction with the movement of the 2nd Armored and 1st Division elements across the front of the Regiment, again pinched out the lines of Combat Team 22.

Activity during August 14th was confined to reconnaissance patrols to the east, which reported indications of an enemy withdrawal to the northeast. In accordance with new boundaries and in view of friendly troops between the enemy and the Twenty-Second Regiment, the rifle battalions were assembled in battalion bivouac areas, and a much needed and well-deserved rest-training period commenced.

The Twenty-Second Infantry had participated in the campaign of the break-through from the 26th of July 1944 to the 14th of August 1944, a total of twenty days. During that time the Regiment travelled ninety-seven miles, and Allied territory in France was increased approximately twelvefold. In the latter part of their campaign, the Twenty-Second Infantry Regiment became part of the force containing the enemy, pending completion of the envelopment move south and east by the Third Army.

The results of the successful break-through of CC 'A' were far-reaching in their effects. Thousands of Allied troops poured through the gap made in the German lines and the entire western German position was upset, greatly aiding the attack of the Third Army along the west coast of the Cotentin Peninsula to Brittany. The Allies finally had sufficient ground and front to maneuver large units. As the situation developed, the Allied Command was able to move powerful armored columns to the enemy flank and rear, and the German withdrawal became a major retreat from Western France.

This campaign, after its initial stages, was characterized by its fluidity and movement, as contrasted to the position warfare of the earlier Sainteny campaign. In place of advancing from hedgerow to hedgerow, there were leap-frogging movements of many miles.

Our losses for the entire campaign were as follows: Six officers and 109 enlisted men killed; thirty officers and 561 enlisted men wounded; and 43 enlisted men missing in action. In another column, 386 German prisoners were captured and processed by the Twenty-Second Infantry. This figure, however, does not accurately indicate the number actually captured, as the rapid movement of the action necessitated that many others be turned over immediately to other units.

The campaign was filled with many notable incidents that will never be forgotten by the men who were there; the all-night move through St. Gilles and burning Canisy shortly after midnight; Captain Reid, Private Sharkey, and the close-in fighting with tanks near Moyen; August 1st, the day on which the last remaining tank of a tank company pushed on with infantry to its objective at Tessy Sur Vire; and the taking of Hills 232 and 230 on August 5th and 6th by the Third Battalion and Company 'C', respectively.

In its first action with elements of a regular Armored Division, the Twenty-Second Infantry Regiment assumed the role of Armored Infantry with unparalleled distinction. General MAURICE ROSE, Commanding General of Combat Command 'A', stated that he never operated with finer infantry than the Twenty-Second.

* * * *

Now that the gap had been opened and troops were able to maneuver, it was known that the next move would be to Paris. It was obvious that the St. Lo breakout had actually broken the main German defenses and commanders believed the next actual encounter with strong hostile forces would not come until the units reached Paris and the Seine River Line. If this was true, the German forces must be kept on the run and not given a chance to form any intermediate lines.

On the 17th of August, after several days' rest, an oral order came down from Division HQ directing CT 22 to move to an assembly area in the proximity of Carrouges, starting at 0800 hours on the 17th, the mission being to re-enforce the 2nd French Armored Division. The Combat Team closed in the new area at 1700 hours and set up the outpost line of resistance.

From the 18th to the 22nd of August, the Regiment remained in place but continued the needed training. During this period the 4th Division passed from the VII Corps to the V Corps, and CT 22 prepared to move tactically by motor convoy to the south of Paris. Shortly before dark on the 23rd of August, the Regimental Combat Team began a motor march to assembly areas near Ablis with the mission of establishing a bridgehead across the Seine River south of Paris. Advance during the night was quick and steady despite a torrential rain. Company 'A', 377th Antiaircraft Artillery Battalion, furnished the column with mobile protection from enemy aircraft.

Shortly after daylight the 24th of August, the Regiment arrived at its assembly area near Ablis. Motors were checked, vehicles were gassed, and at 1200 hours the Regiment moved on to secure positions west of Corbeil preparatory to crossing the

Seine. The march, on an alternate road route, proceeded, and at 2220 hours the Third Battalion was ordered to force a crossing of the river prior to dawn. The Combat Team had covered 109 miles when it closed in this final assembly area.

A change of plans held up this scheme, but shortly after midnight on the 25th, CT 22 was ordered to force the Seine, while CT 12 marched on Paris, and CT 8 reduced a hostile pocket to the west. At 0300 hours, the Third Battalion closed in an assembly area 1500 yards north of Corbeil. Assault boats did not arrive, and only five twenty-man rubber boats were available. At 0630 hours, the Third Battalion attempted to cross the river with four of the boats but were unsuccessful. The four boats were sunk by small arms and antiaircraft fire.

At 1400 hours, the Second Battalion, following a heavy barrage of artillery fire, received the surrender of approximately a platoon of Germans on the far bank. Assault boats arrived simultaneously, and the Second Battalion placed twenty-five men into the hole opened in the German lines by the surrender and progressively widened the bridgehead. By quickly shuttling troops, the entire Second Battalion was across in three hours.

In the meantime, the Third Battalion, 2,500 yards south of the Second Battalion, had succeeded in forcing one company across. The First Battalion moved in between the Second and the Third and also succeeded in getting one company across before dark.

As a result of the above operations, the site for the treadway bridge at Orangis was totally secured by dark. Shortly thereafter, Division received new orders, and the crossing of further units was suspended.

At 2300 hours, the Third Battalion, less Company K and Company C, 70th Tank Battalion, was attached to CT 8 and moved out to join them before daylight.

An improvised barge footbridge across the Seine had been completed by the Engineers, A & P Platoon of the Second Battalion, and members of the Free French of the Interior (FFI) by noon of the 26th. By this time a treadway bridge was also in operation at Orangis, and elements of the Third Armored Division began passing through positions of the regiment. At 1330 hours, the Regiment reassembled on the west bank of the Seine. During the night, the Regiment was alerted for movement the next day to an assembly area in the eastern section of Paris with the mission of driving northeast.

At 0800 hours the 27th of August, Combat Team 22 completed its part in the liberation of Paris by moving motorized into the city. The Third Battalion was relieved from duty with the Eighth Regiment and joined en route. As the Combat Team moved into Paris, they were cheered, kissed, and decorated by the Parisiennes in increasing numbers as the column moved to the center of the city. Time after time, the enthusiastic throngs forced the convoy to halt.

Double Deucers will not forget the move into the city of Paris. Street barricades were swiftly torn down. French cooperation was superb.

The apathy of the French toward the American Army had been most marked in Normandy. The rapidity of the American thrust into central France had prevented any real contact with French civilians prior to this time. American soldiers were hungry for the approbation and friendship of the fabulous Paris, and it was unstintingly given!

French and American flags mingled in the cheering crowds that greeted the Fourth Infantry Division; all Paris opened her heart. French people were everywhere, cheering, laughing, kissing the GI's, patting jeeps, thrusting handfuls of ersatz candy or half-empty bottles of perfume into their hands as they rode by. Paris was free, and so, too, for a time was the soldier. The beauty of Paris lay at his feet; the gratitude of Paris tugged at his heart.

Yet, through the gaiety and the cheering, he was reminded that war was his mistress and fighting his occupation. Standing back from the crowd, under the awning of a sidewalk shop, stood a white-haired French woman. The lines of suffering and grief were deep upon her face, and she was old. No cheering here, and as the Americans rode by, she watched them unsmilingly. Clutched in her hand she held the Croix de Guerre. The ribbon was faded, and we suspected she remembered another war and a son who had given his life for France. The tears rolled unheeded down her cheeks. And we remembered that beyond the cheering and the accolade there remained a German army to defeat, and we were sobered. The blood of our friends and comrades had mingled with that of brave men of other lands. Wounds of the heart can be sharp and deep, and long of healing.

We knew that beyond the borders of France and across Belgium there lay the west wall, Siegfried, mighty warrior, symbol of the hope and armed might of a determined enemy who had first battled for a world and was now at bay, fighting for his own existence.

Thus, the Fourth Infantry Division was first in Paris. It was disconcerting to see newsreels released for American viewing of the French Second Armored Division under General LeClerc given credit for the liberation of Paris. We understood the political and propaganda value, but it still rankled Double Deucers to know the Famous Fourth had been first in Paris, fighting for the liberation of that proud city, only to have it credited to French troops.

Immediately upon arrival at the assembly area, powerful reconnaissance patrols were sent out to the northeast, and the First and Second Battalions prepared to attack abreast. The patrols ascertained that further resistance to the objective was slight.

The scheme of attack was for one company from each of the assault battalions to travel motorized to the vicinity of their respective battalion objectives, then to de-truck and continue afoot. By midnight the objectives were secured. The remainder of the regiment quickly staged forward with the Third Battalion in position to assist either battalion.

Combat in the northeast Paris environ near St. Germain continued on the next day with CT 22 securing a suitable 'jump-off' line for the attack on the 28th of August. Early morning patrols reported that there was indicated enemy activity all along the regimental front. The First Battalion consolidated its positions and tied in with the Second Battalion on its right. Contact with the enemy was general throughout the day. Heavy artillery and mortar concentrations were fired upon located enemy positions. The Third Battalion remained in reserve, so located that it might aid either of the other battalions or Combat Team 8 on the right. During the night, extensive patrolling was carried out.

In compliance with orders, the Combat Team attacked on the left flank of the Fourth Division on the 29th of August. The Second and Third Battalions moved abreast, the Second Battalion on the right. The First Battalion formed the Combat Team reserve. A highly coordinated attack was made, and the objective, Le Mesnil Amelot, was secured by dark. To carry out this five-mile advance, the Regimental Commander previously established a series of intermediate objectives and echeloned the attacking forces to the right.

Negative results were obtained from reconnaissance patrols the night of the 29—30 August. During the day, CT 22 reverted to Division reserve and followed to the left rear of CT 8, protecting the Division left flank.

On 31 August, the regiment was to move via Nanteuil Le Haudouin—Ormoy Villers and to relieve elements of the Fifth Armored Division in securing a bridgehead across the Aisne River. CT 22 completed the motor march of almost nineteen miles without enemy contact, closing in an assembly area in the vicinity of Vez by nightfall.

8. The Race to the Siegfried Line — Through France

With the fall of Paris, it was evident the major portion of the German army would move hastily out of France, organizing its defense behind the vaunted Siegfried Line. Much Nazi propaganda centered around the Siegfried Line. It was termed "formidable," "impregnable" and careful preparations were made by the Allied Command as their armies moved through France toward the German fatherland. Time was a vital factor, and American troops moved swiftly to keep heavy pressure against the German army.

Before dawn on 1 September 1944, the 22nd Combat Team was alerted for a move and shortly thereafter Brig. General George A. Taylor, Assistant Division Commander, arrived at the Regimental Command Post at Chateau Vez.

In two short hours, Task Force 'Taylor' was formed, plans originated, and attachments effected. The missions of the Task Force were to move northward with all possible speed through a designated zone of action, bypassing all points of resistance if possible, and to reach the Corps objective in Belgium east of Valenciennes by dark, 3 September. This distance approximated ninety miles. Because the main bridge at Soissons which was in

VII Corps sector had to be cleared by 1200, the Task Force was forced to move initially with great rapidity.

The first element crossed the IP at 0800 hours on 1 September to follow the route Soissons, Crecy, Guise, Landrecies, to the vicinity of Valencience. A rapid, uncontested march placed Task Force 'Taylor' in Soissons by 1100, Just north of Soissons a deviation was made in the original route so that Combat Command 'Burton' could join the Task Force from a position near Epagny. The First Battalion, led by Major Latimer, Battalion Commander, crossed the highway bridge and then turned left on a dirt trail along the north bank of the river to Pommiers, then turned north to Epahny. General Taylor had prescribed this route with the intention of putting CT 22 in behind Task Force 'Burton' which he knew to be in the general vicinity of Epagny. As a result of this movement, the use of the Pommiers bridge by any other column was effectively blocked.

At 1615 the head of the column met enemy resistance just north of Folembray. Because here there was a narrow road through the woods between banks, the tanks were unable to maneuver. Shortly afterward, Colonel Lanham came to the front and, seeing the situation, immediately returned to the CP at Bonne Maison farm (two miles northeast of Epagny) and obtained General Taylor's approval of a plan to break the Task Force into two columns. Task Force 'Burton,' the Fourth Recon Troops, and First Battalion, Twenty-Second, would be left under the command of Lt. Colonel Ruggles, Executive Officer of the Twenty-Second, and would proceed to the north by the best available route.

The remainder of the force would move by better roads farther east. Due to the confused disposition of the column, which was spread over the various roads between Epagny and Soissons, it was a difficult problem to get underway. Colonel Lanham immediately gave orders to break the column just south of Epagny and just off the highway running north from Soissons at the point where it turned into the secondary road to Epagny; then to have the 747th Tank Battalion, which at that time was on the road south of Pont. St. Mard, take the lead and coil the column in the fields to the east until they had cleared the road south to Epagny and the secondary road running east from Epagny.

When this was accomplished, the 747th Tank Bn. then led the march back down the road to Epagny and the secondary road through Bagnaeus and Juvigny to the highway, thence north up the highway, (the highway bridge over the canal was intact), and to la Fere. The 747th Tank Bn. was followed in order by Second Battalion of the Twenty-Second, Third Battalion of the Twenty-Second, and the rest of the Combat Team. This maneuver proceeded rapidly and without a hitch, and by about 1900 the head of the column had reached the bank of the Oise River of la Fere. There was one point on the road to la Fere where the road went through an archway which was too narrow for the M-10's to pass. This was the only difficulty encountered between Epagny and la Fere, but it caused no delay since the M-10's were immediately run off the road and the column continued.

At la Fere the bridge was out, causing the column to turn southeast to Laon thence north to Crecy. A reconnaissance detachment which had been sent ahead to investigate the bridge at Crecy came under fire from the far bank by a small enemy force, which was dispersed by 2300. However, the bridge was blown, and no other crossing could be found nearby. The column coiled in the fields and waited for the engineers to construct a new bridge.

By dark the west column had eliminated the resistance north of Folembray, taking about 30 prisoners. This column then continued to the Oise River but found the bridge at Chauny out. Lt. Col. Ruggles then turned east along the south bank of the river seeking a crossing. As all bridge crossings to the north were destroyed, the column proceeded in the dark through St. Gobain, Danizy, Reinies, and Pouilly. This column joined the East column just south of Crecy at dawn, the 2nd of September.

By about 0830, September 2nd, the new bridge had been constructed by the engineers and the Task Force resumed the advance to the north in a single column. Only occasional slight contact with the enemy occurred until the Sambre River was reached. There was a brief halt near Foucouxy while tank destroyers were brought up to knock out a German Mark V tank; several German vehicles were destroyed at le Herie; and a wagon column was destroyed a mile north of Guise. At Iron River, the column again split, Lt. Col. Ruggles taking Task Force "Burton" and the 1st Battalion, 22nd and turning west on the road to le Cateau. This column had a skirmish at Hannapes where they secured the Sambre Canal bridge before it could be destroyed, even though it had been prepared for demolition. This force met other small enemy forces near Wassigny and Ribeauville.

The east column continued to Landrecies, meeting only occasional small groups of the enemy. At Landrecies the column found the main bridge over the Sambre River destroyed. (The bridge had been completely blown up with unusually heavy charges.) Within 30 minutes an old railroad bridge (trestle) was improved sufficiently to be used. The head of the two columns crossed and continued several miles north of Landrecies and surprised an enemy column, inflicting extremely heavy losses.

Early in the evening on September the 2nd, orders were received to halt the advance. The east column was to bivouac on the south side of the canal, excepting the 747th Tank Bn., Company

K, Twenty-Second Infantry, and the 4th Recon. Troop, which had already crossed. These latter units held a defensive position around Landrecies to protect the crossing. This position was out-posted by the 4th Recon Troop which, during the night, contacted the enemy apparently trying to move to the east. The Recon outpost knocked out two German sedans and a self-propelled gun. The west column bivouacked for the night around the cross-road just north of Ribeauville.

During the advance on 2 September, the Task Force had been moving on one of the main routes of retreat of the enemy. In addition to occasional German vehicles overtaken and destroyed on the road, many small groups of German soldiers fled into the fields as our column came by. In some cases, these groups fired on the column, in others the Germans surrendered. A considerable number of prisoners were picked up in this manner during the day and were turned over to the Free French of the Interior.

During the night an order was received to move the Task Force into an assembly area in the vicinity of Pommereuil, be-tween Landrecies and la Cateau. At 0800 hours on 3 September, the east column moved toward Pommereuil. Our planes could be seen attacking something on the road not far ahead. When the column reached it, this proved to be a German column consist-ing of many more assorted motor vehicles and at least 40 horse-drawn vehicles, with infantry of about a regiment strength. Many of the vehicles had already been destroyed by the air attack; the remainder were knocked out by the Task Force. The German in-fantry of the column had deployed and put up a fight which was not finished until sometime in the afternoon. When it was over, some 200 or 300 prisoners had been taken and the rest of the enemy killed. There was so much destroyed enemy equipment on the road that it was necessary to use a bulldozer to clear it.

Meantime, the west column had not received the order to assemble until about noon. They then moved to Pommereuil, passing southeast of la Cateau. By 1700 the entire Task Force had closed in the assembly area. The area Landrecies — le Cateau — Poix was mopped up during that night and the next two days. The Recon Troop was given the mission of cleaning out the western portion of Foret de Normal. Several hundred prisoners were taken during the night and still more were taken throughout the next two days.

The Task Force processed 1,300 prisoners taken in the le Cateau-Landrecies area in addition to several hundred which had been taken on the march and turned over then to the FFI. The enemy was evidently greatly confused in the le Cateau — Landrecies area and during the night of 3 September there were cases in which Germans started digging in close by our units.

General Taylor had a bullet pass through the side of his jacket, just grazing his skin, while he was standing on the railroad bridge at Landrecies. This same bullet had first passed through Colonel Lanham's sleeve.

At la Groise when the column halted, a gasoline truck, methodically keeping his 50-yard distance, came to a stop in the middle of a road junction. Within a few minutes an enemy gun a considerable distance away on a side road fired one round which hit the truck squarely and blew it up, killing the driver and another man. A few minutes later, while the remains of the gasoline truck were burning fiercely, a terrific explosion occurred; the 22nd Infantry had blown up a German ammunition dump nearby, but on the explosion everyone ducked for cover. A few minutes later a Frenchman dashed out of a house near the road junction, snatched the tri-color, which was stuck in the fence, and disappeared into the house with it.

At 0300 hours the 4th of September, a warning order was received from Headquarters 4th Infantry Division to alert a reinforced battalion to move at daylight to Brunehamel and report to the 102nd Cavalry Group, to which it was to become attached upon arrival. The 1st Battalion, commanded by Lt. Col. John Dowdy, was the designated unit and they moved out in the early dawn. Later that morning, the Task Force was disbanded. The balance of the Regiment continued mopping up operations that day and through the night.

9. The Race To The Siegfried Line — Through Belgium and The Ardennes

THE MYTH EXPLODED —
"FESTUNG GERMANA"

"His name from me let him take — 'Siegfried'; for Sieg-fried in TRIUMPH shall live!"

— Sieglinde

"…With blood were all bedabbled the flowerets of the fields. Some time with death he struggled as though he scorned to yield."

(From Carlyle's translation of Fragments of the poem)

Following up the northward drive, which had driven the Germans well back, the units prepared to push on through Belgium and into the Siegfried Line. Everyone believed that now it would be only a matter of a few weeks before the retreating German armies would be forced to surrender. Things were going well for the 22nd Regiment — remarkable gains had been made since Paris, and casualties were almost nil.

At night it was still possible for artillery flashes to be seen from the positions of the 44th Field Artillery Battalion supporting the 22nd. Since leaving Paris, however, the 44th had fired only fifty rounds.

On the morning of September 7th, movement orders were received from 4th Division Headquarters for the Combat Team to rejoin the Division in Belgium. Leading elements crossed the IP south of Pommereuil before noon. In accordance with orders, the Combat Team proceeded to Graide, Belgium, some eighty-five miles, closing in before dark. Upon arrival, the Combat Team was placed in Division reserve to secure the flanks of the Division with patrols. These past few days had been relatively easy, and the officers and men were enjoying a period of appreciated relaxation.

On the afternoon of September 9th, the 3rd Battalion attacked between the 8th and 12th Regiments and, despite some enemy resistance was able to move nearly seventeen miles east to Giuroulle.

Because the Allied forces were advancing so rapidly, supplies were becoming scarce. It was almost impossible for the service troops in the rear areas to keep the supply trains up with the leading elements. Roads were torn up, and bridges were out. Quartermaster trucking companies worked day and night in an effort to get the supplies up to the units. Therefore, the greater part of the supplies were being flown in by transport planes and dropped to the troops. One of the most serious shortages was in the available gasoline supply, and, as a result, trucks had to be dispatched to move only one battalion at a time.

The morning of the 10th, the 2nd Battalion moved by truck to the rear of the 3rd Battalion at Saile and de-trucked. The trucks then returned for the 1st Battalion. In the meantime, the 3rd Battalion pushed east through Gives and Compogne and on to Houffalize. When the head of the column reached the southern outskirts of Houffalize, the road was found to be blocked by fallen trees and covered by automatic weapons. While the Engineers were clearing the road, the Regimental I and R platoon entered the town by a narrow trail paralleling the main highway. As soon as they had entered, a guide was sent back to lead the remainder of the column into the city. The citizens of Houffalize were apparently overjoyed at the sight of the American soldiers and did everything possible to help. They were materially responsible for the rapid progress of the Combat Team.

They assisted in clearing the roadblocks and later rebuilt a destroyed bridge so that it was capable of withstanding the weight of the heaviest vehicle. The citizens readily reported the disposition of, and the size of, the enemy forces as they had been several days before. That evening the town was cleared. The 3rd Battalion moved a short distance northeast and the 1st and 2nd Battalions closed into areas south of Houffalize.

The advance pushed on again the next morning. Despite roadblocks and enemy small arms and artillery fire, the Combat Team was able to secure its objective, Beho, by dark.

Late that night, a strong reconnaissance patrol was organized with the mission of crossing the German frontier. 1st Lt. ROBERT L. MANNING, 3rd Battalion Scout and Raider Platoon Leader, was selected to lead the patrol. In addition to the scouts and raiders, the patrol had attached to it two self-propelled tank destroyers and five jeeps. Actually, the patrol route extended some eight miles. The patrol was charged in addition to obtaining enemy information, to return with a jar of German soil which was to be sent to the President of the United States. When the patrol reached the Our River, it was impossible for the vehicles to continue, and the patrol was forced to continue on foot. Because of this, Lt. MANNING and Lt. SHUGART flipped a coin to see who was to continue with the patrol. Lt. C. M. SHUGART, the I and R platoon leader, won the toss and led the foot patrol into Germany. This patrol is believed to have been the first organized allied unit to cross onto German soil during World War II. It crossed the border near Hermmeres, Germany, at 2130 hours, the 11th of September 1944. No casualties were sustained, and the mission was a success.

On the 12th, the advance was resumed, led by a strong combat patrol commanded by Lt. E. C. MARTIN. The patrol consisted of three tanks, two tank destroyers, six jeeps from the Scouts and Raiders of the 1st Battalion, and one rifle platoon led by Lt. RUBACK of Company "A", all of which were attached to the 3rd Battalion. Another patrol went out at the same time, commanded by Lt. M. W. TOLLES and worked in conjunction with the 2nd Battalion. The patrols crossed the German border at noon and engaged in numerous skirmishes with the enemy, destroying one enemy tank and damaging another. The patrols upon return reported that the resistance encountered had been light. The patrol had captured three prisoners, killed one German, and several had escaped wounded. They had penetrated to the high ground beyond Elcherath and then had moved two miles north along the highway.

At 0829 hours, September 12, Battery 'C' of the 44th Field Artillery Battalion fired what is believed to have been the first light artillery shells to strike Germany proper.

At 0845 hours, orders were received for the entire Combat Team to move to the high ground just west of the German border. The 2nd and 3rd Battalions moved forward online, the 3rd on the left (north). The 1st Battalion followed up in trucks. Later, Division ordered movement continued, and by 1500 hours, the northern column was two miles east of Gruflange and the southern column was at Burg-Reuland. The leading elements of the 3rd Battalion crossed the border in the vicinity of Hemmeres, Germany, at 1617 hours.

At 1700 hours, the Combat Team Commander was notified that Major General R. O. BARTON, Commanding General of the 4th Infantry Division, was on his way to the advance command post. The Commanding General arrived at the Regimental CP at 1715 hours, the first American Division Commander to enter Germany.

A strong Combat patrol led by Lt. MARTIN, without tank support, moved forth again on the morning of the 13th. The patrol went to Bleialf where they first observed the Siegfried Line. From here they went to Buchet, encountering only light resistance. All enemy contact was scattered and disorganized. At 1015, the Regiment followed the route of the patrol. Opposition was slight, and by 1130 hours, the 3rd Battalion was just east of Urb, and the 2nd Battalion was at Groslangenfeld. An hour and a half later, both battalions were at their objectives in the vicinity of Bleialf. Company L, which was following closely behind the patrol, moved on to Buchet, upon recommendation of the patrol leader, and by night the town was secure.

10. We Enter Germany First

The first day the Combat Team entered Germany (12 September 1944) they captured the towns of Hemmeres, Steffhausen, Auel, and Elcherath, the latter the site of the first Regimental Command Post on German soil. On the 13th, the towns of Winterscheid, Bleialf, Buchet, and Schweiler were added to the list. Civilians did not interfere, but instead just stared in amazement at the advancing columns and hung white flags on all the buildings.

The Combat Team was now face to face with the Impregnable West-Wall. This is the name which the Germans gave the Siegfried Line, as it actually formed a wall along the western border of Germany. The basic principle underlying the construction of the Siegfried Line was that of diffusing rather than concentrating Germany's far-flung defensive system which extended in depth from France to Holland.

When Adolph Hitler first decided to construct such a defensive line, he encountered the problem of convincing his staff that a diffused system was better than an intercommunicated system. To emphasize his theory to his general staff officers, he used this comparison. He placed a China platter beside a ten-pfennig piece and asked them, "Gentlemen, which is harder to hit—the plate or the coin?" To which they obviously replied, "The coin."

This illustrates one of the basic differences between the French Maginot and the German Siegfried fortification —

- Maginot: one gigantic, continuous, and interconnected system.
- Siegfried: innumerable smaller units, of which one or the other might be captured or destroyed without endangering or weakening the rest.

German tactics in constructing the Siegfried Line centered on three general ideas: How can bombing attacks be made innocuous? How can tanks be prevented from invading this zone? Can artillery fire be so diffused as to make big gaps in the Westwall impossible? The Germans actually constructed different types of fortifications and then put them to a test with the above questions in mind.

In certain sectors, the Siegfried defenses extended in depth as much as thirty-five miles from the frontier and consisted of a complicated maze of barbed wire and crisscrossed concrete and steel constructions, strewn on the soil like so many dragon's teeth; pillboxes and bunkers camouflaged into the surrounding landscape; and fortress-like, armor-plated dugouts invisible to the human eye, and connected with numerous subterranean passages. The system is generally four lines of defense — the first two for infantry and artillery, the third and fourth for anti-aircraft.

Since these facts had been known and studied by strategists for several years, Colonel Lanham, Combat Team Commander, realized that the sooner he struck the line with his forces, the better the chance of penetration. So as soon as the troops were against the Line, he held a meeting with all battalion commanders, separate unit commanders, and attached unit commanders. The meeting was held in Schweiler, Germany, the night of September 13th. Plans were formulated, attachments made, and shortly after midnight, orders were issued for the initial assault on the Siegfried Line at 1000 hours, the 14th of September.

The scheme of maneuver was to be an assault in a column of battalions in the order Third, First and then the Second. The Third Battalion was to have attached to it Company C, 70th Tank Battalion, Company A, 893rd Tank Destroyer (SP) Battalion, Company A, 81st Chemical Battalion, Demolition Detachment, Company C, Fourth Engineers Battalion, Company A, Twenty-Second Infantry, and the Mine platoon of Antitank Company. This battalion, with attachments, was then to move to an assembly area near Buchet and attack, penetrating the line about 1,000 yards east of Buchet to the main northeast-southwest road directly to the rear of the first fortifications. The Third Battalion was to destroy all fortifications enroute to its objective, the high ground approximately 600 yards south of the initial penetration.

One platoon of tank destroyers and the company of tanks (less one platoon) were sent to the left at the initial penetration area to cover the gap protecting the north flank. The First Battalion, plus one platoon from Antitank Company, was to advance via Bleialf and, following the Third Battalion by 750 yards, pass through the gap made by that unit. This battalion was also to pick up the platoon of tank destroyers, Company A, and the tanks left in the gap by the Third Battalion, and push north 1,500 yards, reducing all fortifications in its wake to the Combat Team boundary. The Second Battalion, originally in reserve was to protect the south flank and be prepared to stage forward and move against any counterattack.

At 1130 hours on the 14th of September, 1944, the attack took form. The 3rd Battalion leading, the Combat Team reached the Siegfried Line at 1305 hours, penetrated, and cut two lateral roads in the rear of the first fortifications approximately 900 yards east of Buchet. This penetration was made against a hail of small arms, intense mortar, and artillery fire, and fierce antitank fire from 88's. Numerous pillboxes were located which were not indicated on intelligence maps or photos. The enemy, partially SS troops, refused to surrender until wounded or blasted from his fortification. The First Battalion attacked, as had been planned, and pushed through the gap previously opened by the Third Battalion and then moved north against stubborn resistance. The Second Battalion also pushed through the gap and by 2100 hours had taken its position. Throughout the day, the battalion fought a close-in hand-to-hand battle. Casualties were moderate, some received from heavy anti-personnel mines. In comparison, the Germans sustained much heavier losses, but our Capt. Reid was killed September 14—a great soldier.

During the afternoon, orders were received from Division changing the southern boundary to include Brandscheid, which had previously been in the 28th Division sector. When the troops dug in for the night, the 3rd Battalion was in position to attack Brandscheid and the remainder of the Combat Team was prepared to continue the attack to the east.

The next morning a very serious enemy infiltration delayed the planned attack of the 3rd Battalion on Brandscheid for several hours. Overcoming this obstacle by noon, the battalion slowly pushed south against determined foes. Shortly thereafter, an enemy force estimated to be at least three hundred in number made a raging counterattack against the 2nd Battalion, partially surrounding them and demanding a surrender. Within a few hours the counterattack was repulsed, and the enemy had suffered heavily in casualties.

The 3rd Battalion was able to advance very little and spent a great amount of time clearing the pillboxes near the crossroads at Meisert. Late that same afternoon, the Battalion consolidated its gains and prepared to continue the attack for Brandscheid the following morning. Enemy resistance was definitely on the increase as more counterattacks were made, infiltration increased, and artillery pounded the front line troops the entire day. Lt. Col. W. A. WATSON, Commanding the 44th Field Artillery Battalion, remarked that the enemy counter-battery artillery fire was the greatest encountered since the Cherbourg fighting.

The Combat Team quickly dug in for the night as the enemy artillery continued to pound them. The burst from both mortars and artillery shells took casualties in mounting numbers. As many men as possible crowded into the captured pillboxes for the night. Now as before, during any engagement fifty percent of the command must be awake at all times. It was almost an impossible task for already exhausted troops to have to remain awake on guard. Only the green replacements had to be watched to be sure that they were alert. The men who had seen the results of an unalert guard were aware of the constant vigil needed to locate enemy patrols moving stealthily through the darkness.

The 1st Battalion attacked late on the morning of September 16th to seize the high ground west of Sellerich, with the 2nd and 3rd Battalions in the gap. In the face of the heavy artillery, mortar, and small arms fire, strongly defended bunkers were taken. It was in this drive that the First Battalion Commander, Lt. Colonel John Dowdy, was killed. One of the finest officers of the regiment, Lt. Colonel John Dowdy had proved himself an able officer in combat again and again. His personal care for the troops under his command, his knowledge of military tactics, had saved lives and boosted morale. The 3rd Battalion, delayed for a time by rain, made slow but steady progress, and by 1315 hours the attack was moving, and one company reached the crossroads within the town of Brandscheid. The attack again was delayed when the by-passed pillboxes had to be cleared. This was done, and positions were secure by nightfall. The Combat Team had suffered several casualties in the day's operation.

All battalions, re-enforced by tanks and tank destroyers, jumped in the attack on the morning of September 17th. In spite of the heavy enemy artillery, the 3rd Battalion was in Brand-scheid by 1315 hours and had captured several Germans.

While this action was going on in Brandscheid, the First Battalion had encountered serious opposition. Major Robert B. Latimer, who had assumed command of the battalion after Colonel Dowdy was killed, planned to take his battalion by way of the draw north of Sellerich to Hill 520, his objective. The other route of approach would be under observed fire from three directions. The enemy held the hilltops to the north, east, and south of Hill 520.

Company A advanced down the draw to the east and across the north-south draw, still under heavy shelling but meeting no small arms fire. No enemy were found in the village of Sellerich, and nothing but the enemy artillery opposed Company A's advance. They found no gun positions within their zone of advance. (It turned out that the enemy guns were 500 to 1,000 yards away on each flank!) All three platoons of Company A and the attached machine gun platoon of Company D moved on to the top of Hill 520 where they are believed to have occupied a front of four or five hundred yards.

Company B with the tanks followed Company A into the low ground between Hontheim and Sellerich. At this point, the enemy sprang its trap. (A prisoner later stated that it was a deliberate trap, that the Germans had watched Company A go up the hill and deliberately held their fire in order to let them get far enough forward to cut them off). As Company B and the tanks got into the north-south draw, one of the tank destroyers hit a mine which knocked off a track. Then enemy AT guns firing along the draw from both north and south drove the tanks back every time they attempted to cross the draw. At the same time, the enemy opened violent fire with weapons of all sorts from both flanks. Flak guns, AT guns, and machine guns were massed on the hillsides northeast of Hontheim and south of Herscheid, enfilading the north-south draw from both ends. This fire completely stopped Company B and the tanks, pinning them down in the low ground between Hontheim and Sellerich.

At the same time, the enemy put down very heavy artillery fire both on Company B and on A on top of the hill. Company A had not had time to finish digging in and suffered heavily from this fire. German infantry also opened fire from entrenched positions (including civilian air raid shelters) in Hontheim and Herscheid.

Companies A and B were suffering severely in casualties and were later told to withdraw to positions held the night before. Company K, together with the tanks previously attached to the 3rd Battalion, went to the aid of the 1st Battalion. Lt. Colonel John F. Ruggles, Regimental Executive Officer, led Company K to the 1st Battalion CP and coordinated the relief. Lt. Surratt assumed command of Company A after Lt. Marco was wounded and was able to partially reorganize the company, even though casualties were over fifty percent. By dark, the 1st Battalion again occupied the positions held earlier that day.

Because the 3rd Battalion temporarily lost men and tanks, it was not able to continue the mopping up of Brandscheid and was forced to re-adjust its lines near Meisert. The 2nd Battalion reached and secured the northern half of its objective. Active security patrols were sent out that night, and the Regimental I and R platoon screened the gap between the 1st and 2nd Battalions—an enormous job.

At 2000 hours, orders came from Headquarters Fourth Infantry Division to cease any further attacks on Brandscheid. The mission of the Combat Team now was to improve, maintain, and secure its position in the vicinity of Buchet, patrolling to Sellerich, Hontheim and Brandscheid.

From the 18th of September to the 3rd of October, the Combat Team remained in a static position. Lines were secured and adjusted. Enemy counterattacks were numerous but always repulsed. The Regiment sent out vigorous patrols day and night to keep in continual contact with enemy activities. Enemy movement was reported to be extensive around Hontheim, but never materialized. The 44th Field Artillery Bn. in the days of the holding action had fired over 15,000 rounds in support of Combat Team 22.

The assault and penetration of the Siegfried Line revealed several important facts.

In the assault of the line of fortifications, the best method was found to be an attack, in strength, on a relatively narrow front. Once the line was pierced and the pillbox system of mutual support disrupted, it was fairly simple to roll back the line, hitting the boxes from the flanks or rear. At this stage of the operation, the greatest difficulty was fending off attacks by small enemy groups issuing forth from rear positions or by-passed emplacements. The excellent camouflage of the pillboxes and presence of more than shown on G-2 maps caused numerous emplacements to be by-passed, and it was necessary to form patrols to search them out.

The majority of pillboxes (estimated at 75%) were destroyed by the coordinated attack of direct fire weapons: a tank, a tank destroyer, and infantry. This method was used whenever terrain permitted. Supporting fires from any other available direct fire weapons, including .50 cal. machineguns and 57mm antitank guns, were used to chase the enemy into the pillbox and force the closing of the firing embrasures or door. Under this cover, a tank destroyer with close-in infantry protection would move to within 15 or 20 yards of the box. From this close range the tank destroyer would preferably blow out an embrasure or door. If these could not be reached, the three-inch guns would pierce the concrete sidewall itself. The infantry then moved in, mopping up with fragmentation and white phosphorous grenades, or if necessary, with demolition charges and flame-throwers. Flame throwers were very effective when fired through holes blown by the tank destroyers. Experience proved that best results were obtained by first squirting some of the liquid, not ignited, into the hole, and following that immediately with a burning blast.

When terrain and trees would not permit the close approach to a pillbox by a tank destroyer, tank-infantry tactics were used. Tank fire together with any other available direct fire weapons, forced the enemy into the box. If fire could be brought to bear on the embrasures, they were of necessity closed. At least the enemy's field of fire was greatly restricted by the embrasures. The infantry then closed in, and with flame-throwers and demolition charges cleaned out the fortifications. Pack charges effectively blew out embrasures or doors, and the concussion either killed or stunned the occupants. The pillbox could then be mopped up with grenades or flame-throwers. An infantry combination of one assault (demolition) squad and one support (rifle) squad, worked well. Lack of men trained in demolitions and flame-throwers handicapped this method. In an infantry assault on a pillbox, captured German bazookas (hand-panzer) were very effective. These weapons would knock a hole through the concrete, and the concussion and blast effect killed the occupants. Our bazookas were not effective.

In some instances, a tank dozer was used to seal pillboxes. Sunken doorways were easily blocked in a similar manner, although this was more difficult, the embrasures being above ground level. This method prevented occupants from escaping as the doors, which swing out, were blocked. However, several boxes so buried were excavated from the outside and reoccupied by the Germans.

Ernest Hemingway with Colonel Charles T. Lanham by a captured German flak gun at the Siegfried Line in Germany September 18, 1944 — Photo from the National Archives NARA 192699

11. The Siegfried Rests

The Twenty-Second Infantry had spent slightly less than three weeks inside the Siegfried Line. Though the regiment had not actually breached the Line, they had penetrated it to a depth of several miles. Had the necessary supplies, ammunition, and equipment been available, it is believed that the Line could have been breached and the Allied forces driven well into Germany. At this time, the German armies were staggering and confused to such an extent that weeks would have been required before they could have organized a defensive line.

On October 4, 1944, the Combat Team was ordered to move to a new area near Bullingen. Elements of the Second Division began relieving the Twenty-Second Infantry on October 3, and it was completed on the 4th without enemy interference. The 1st Battalion and special units, immediately upon completion of their relief, moved to an assembly area near Honsfeld.

Early on the morning of the 5th, the 1st Battalion initiated reconnaissance to affect the relief of the 102nd Cavalry Reconnaissance Squadron in the Combat Team's new zone of action. This completed, the battalion moved into position immediately to the rear of the cavalry outpost line. That part of Combat Team 22 which remained in assembly areas near Buchet moved north by motor convoy on the 5th and arrived in the new area at Honsfeld that afternoon.

Strong security patrols were maintained on the 6th of October in order to protect the relief and aid the oncoming attack. The attack moved out in an approach march formation before noon that day, and by 1400 hours the leading elements of the 1st and 2nd Battalions were 1700 yards beyond the line of departure. The 3rd Battalion, following up in reserve, moved into the vicinity of Murringen. Contact was continuously maintained with Combat Team 112 on the left and Combat Team 8 on the right. Combat and reconnaissance patrols from the two forward battalions which were sent out short distances east of our lines that night reported that they had seen entrenched enemy within sight of our positions.

Even though the Combat Team had shifted its zone of action to the north, it was still located in the Schnee Eifel and Monschau Forest area. There was, in reality, very little aggressive action aside from the skirmishes encountered by the combat patrols.

From October 8th to the 10th, patrol activity continued in an effort to keep in close contact with enemy plans and movements. A coordinated plan was drawn up for the attack on the Siegfried Line. The plan was never put into effect since orders were rescinded, and the mission of the Division was changed.

Before the official delay of plans arrived, the field artillery battalions, in support of CT 22, opened fire on the afternoon of the 11th of October and fired a number of heavy concentrations. Additional support was given by the IX Tactical Air Command, which dive-bombed and strafed enemy positions in the vicinity of Udenbreth. As a result, partial withdrawal of enemy troops in the forward positions was observed. When information was received to delay the attack, plans were set up at once for out-posting the Combat Team with one battalion — two battalions to be in reserve.

From the 12th of October until the 22nd of October, the Combat Team remained relatively static. Patrols were constantly on the move reconnoitering to the east to include the towns of Miescheid and Udenbreth. Only slight casualties were sustained. The weather increased in severity, and the countryside was under an oppressive overcast of drizzling rain and penetrating cold. Every advantage was taken of the comforts that could be enjoyed. Showers, movies, Red Cross facilities, USO shows, were setup in Murringen, and daily quotas of men were sent back from the reserve units to enjoy them while they were available.

Relief of the 28th Infantry Division began at 0700 hours on the 23rd of October. The battalion's side-slipped to the north in the order 1st, 3rd, and then the 2nd. The 1st and 3rd Battalions, Combat Team 22, occupied that front previously held by Combat Teams 109 and 112, 28th Infantry Division. The relief was completed by early afternoon, and the battalions established and improved the positions to the northeast and coordinated their defensive fires. This new position placed the defending battalions on a front previously covered by an entire division.

Through the remainder of October and until the 5th of November, the Regiment occupied a position with its front paralleling the Siegfried Line fortifications in the proximity of Krinkelt, Belgium. With no indication of orders for an attack, the Combat Team confined its combat activity to harassing fires and patrolling. The period was highlighted by sharp patrol clashes, considerable harassing and interdicting artillery and mortar fire, and harassing long-range machine-gun, 57mm antitank, and self-propelled weapons fire. The battalions, whenever in reserve, conducted training on demolitions and mines, and physical training.

The relief of the Combat Team by the 39th Infantry, 9th Infantry Division began at 0800 hours the 5th of November. The Combat Team was fully relieved by 1500 hours and moved to the previously selected assembly area at Krinkelt and dug in. Staff planning was instituted immediately for a movement to an assembly area near Zweifall, Germany. Earlier that morning, a quartering party of officers and enlisted men had moved to that sector.

On the 7th and 8th of November, the Regiment carried out various forms of training, physical reconditioning, and weapons firing. A fifty vehicle detachment was sent ahead with the mission of making a tactical reconnaissance of the new sector. Rigid security regulations designed to preserve the secrecy of the Combat Teams' move were instituted. Division shoulder patches were removed or covered, and the move was to be made at night.

The Twenty-Second Infantry Regiment was in top peak; it was a fighting machine trained to an efficiency not matched at any other time during the war. It was an aggressive, battle-scarred, confidently experienced regiment bent on the destruction of the enemy.

The landing at Utah Beach, the St. Lo Breakthrough, the Battle of Paris, and the initial penetration of the Siegfried Line all stood as major achievements of the Twenty-Second Infantry Regiment. Perhaps no Regiment of the United States Army ever had a more impressive battle record in such a short space of time. The Regiment was poised for battle; it could not know the shattering hell of the eighteen days ahead.

Soldiers of the 22nd Infantry entering a village in Belgium October 1944 — U.S. Army Signal Corps photo #ETO-HQ-44-14676

12. The Death Factory — Hurtgen

"Show me a man who went through the Battle of Hurtgen Forest and who says he never had a feeling of fear, and I'll show you a liar or a damn fool. You can't get all of the dead because you can't find them, and they stay there to remind the guys advancing as to what might hit them. You can't get protection. You can't see. You can't get fields of fire. Artillery slashes the trees like a scythe. Everything is tangled. You can scarcely walk. Everybody is cold and wet, and the mixture of cold rain and sleet keeps falling. Then we attack again and soon there is only a handful of old men left."

—T/5 George Morgan, 1st Bn.

The following narration concerns only one of the battles fought by the Twenty-Second Infantry, but even though it lasted only eighteen days, the percentage of casualties far exceeds those sustained at any other time, including the invasion of Europe.

Many of the combat veterans who fought through the D-Day landing and later the Hurtgen Forest, ranked the Hurtgen as by far the bloodiest, most filthy fight they had encountered. They knew then they were really the front-line riflemen. It was in such a battle as this that the true heroism of the infantry doughboy came forth. There is no other branch of the service where the men must eat, fight, and live in the mud. These heroic men fought continuously within fifty yards of the enemy, often with actual physical contact and with sure death only seconds away. These men ate the issue rations when they were frozen, muddy, and stale. Fires were unheard of. These men lived day and night in the bloody slime to be found only in the Hurtgen Forest. It was not an uncommon sight to see a dead soldier with the pit of his stomach ripped open, with his head blown completely off, with his back broken by shrapnel, or to hear the wounded scream in terrifying pain with their legs or arms completely blown off by an enemy shell. The Forest itself is an indescribable scene which holds within its depths the horrifying memories of each individual who fought there. The men of the 44th Field Artillery Battalion supporting the Twenty-Second know what living hell is, for they followed closely behind the rifle troops in order to better protect them.

As the account of the Hurtgen Forest is uncovered, one must keep in mind the horrible surroundings in order to appreciate the gallant struggle put forth by the determined infantry units.

The Combat Team moved from its positions along the Siegfried Line to an assembly area near Zweifall, Germany, on the 9th of November 1944. When the Combat Team closed in at Zweifall, it was assembled in the western fringes of the hilly, thickly wooded Hurtgen Forest. Immediately, the troops constructed wooden dugouts for protection from the constant sleet and from enemy artillery.

Between the 10th and 15th of the month, the Regiment continued its preparations for the oncoming offensive. The weather continued to be miserably cold and damp. Schools were held for all company grade officers in woods fighting, map reading, and the adjustment of artillery fire. Too, all enlisted men and officers were told how to make a 'shell rip report' on enemy artillery, a method of determining range, size, and azimuth of the piece that was firing. Anticipation of difficulties resulted in intense work with communication teams, company aid men, litter bearers, and 81mm mortar platoons.

In the initial phase of the attack, the three 81mm mortar platoons were to be massed as a regimental unit with telephone communications to each battalion CP. Lt. Eidson, from D Company, was to coordinate the three platoons in the fire missions. Because the 2nd Battalion was to primarily be the assault unit, priority of fire was initially to be given to them, then shifted where needed most.

At 0100 hours, CT 22 was notified that D-Day was to be the 16th of November and H-hour to be 1245 hours. The awaited assault on the Hurtgen Forest was about to begin.

The plan was for the attack to be a column of battalions in order 2nd, 1st, 3rd. The 44th Field Artillery, re-enforced by the fire of the 20th Field Artillery, was in direct support. Attached to the CT were medium tanks of Company C, 70th Tank Battalion, light tanks of one platoon of Company D, 70th Tank Battalion, 4.2 chemical mortars of Company C, 87th Chemical Battalion, tank destroyers from one platoon of Company C, 803rd Tank Destroyer Battalion SP, one platoon of Company C, Fourth Engineer Battalion, and Company C, Fourth Medical Battalion. Due to the dense woods and rugged terrain, the armor and tank destroyers were initially to be in reserve. The mission of the CT was to drive the Germans from their defensive positions in the Hurtgen Forest, the northern limit of the Siegfried Line fortifications, and open a path for a coordinated infantry-armor thrust across the plains of Duren and later the Rhine.

The Hurtgen Forest that lay ahead was, in a matter of only a few days, to become one of the bloodiest battlefields in the world. The trees and undergrowth of the forest were so dense that areas within the forest had never seen sunlight. The trees were of various types—fir, spruce, pine, and cedar. Roads within the forest were few. The only means of moving about was by means of fire breaks, lanes cut through the timber, which were extremely narrow and poorly kept. Actually, Hurtgen was a cold, forest hell—a death factory. It blocked the main approaches to Cologne and the Ruhr Valley; Hurtgen was a 'must' objective. The terrain was difficult enough, steep hills, poor roads, numerous creeks, deep canyons, and thick woods. Across the front stretched belts of mines and barbed wire rigged with thousands of booby traps. Dug-in machine guns and automatic weapons were placed to spray the entire area with interlocking fire. Artillery, doubly dangerous in the woods because of tree bursts, was zeroed on every conceivable objective, road, and fire break.

Weather, a highly important factor, was pure misery. Constant rain, snow, and freezing temperatures were continuously hampering operations. Living for days in water filled holes, usually without blankets and never a fire, troops had no escape from cold and wet.

Before the main offensive got underway, heavy artillery preparations were fired to the right (south) and to the left (north) flanks of the planned penetration area, as a planned means of deception. At 1245 hours the 2nd Battalion crossed the line of departure, a road running generally northeast-southeast and approximately 5,600 yards northeast of Zweifall. They advanced over 600 yards before enemy contact was made. The battalion continued to advance against slight resistance, dropping one rifle company to form a defensive flank to the north. Fifteen hundred yards from the line of departure, the battalion met heavy mortar and artillery fire, which forced them to halt and dig in for the night.

The 1st Battalion following the 2nd, turned north, passed

through the 2nd Battalion rifle company, which had been dropped off to protect the north flank, and attacked north along Trail E. This drive struck the flank and rear of an entrenched enemy battalion. As the attack progressed, resistance increased. As soon as favorable terrain was secured for the attack the next morning, the battalions dug in. At 1500 hours, the 3rd Battalion, temporarily lagging behind, sent one company to relieve elements of the 2nd Battalion holding the southern portion of Trail E. Another company followed, and as the relief was completed, they both moved to the southwest of the 2nd Battalion to protect the southern flank of the CT and to further protect Trail E, the temporary supply route. This trail had already been cut on the north by the 1st Battalion advancing toward trail fork X.

As each battalion dug in for the night, new orders for the next day arrived. It was almost impossible for plans to be set forth more than twenty-four hours in advance. As reconnaissance patrols were never able to extend more than a few hundred yards in front of the lines, troop leaders rarely knew the exact disposition of hostile fire power until they actually exposed themselves. As a result, intermediate objectives were assigned to the battalions each night. As the troops dug in for the night, Captain Henley, 1st Battalion Executive Officer, aptly stated, "A man would throw away his rifle before he would give up his ax, because with an ax he could chop wood and also kill Heinies." This was readily apparent because every man who could obtain a pick, an ax, or a shovel carried it as he did his weapon.

The scheme for the 17th called for the 1st Battalion to continue its advance along Trail E and seize the hill 1,000 yards north of there and trail junction X on its forward slope. When this was accomplished, the battalion was to swing east and attack online with the 2nd Battalion to seize the high ground approximately 500 yards to the front which controlled Road A, the main north-south road through the forest.

Because the enemy laid down extremely heavy mortar and artillery barrages in the Combat Team sector, the attack was delayed over an hour. During the barrage the 1st Battalion Commander, Major Hubert L. Drake, was killed and the 3rd Battalion Commander, Lt. Colonel Arthur S. Teague, was wounded. Major Goforth, who was at the Regimental CP at the time the news was received of the death of Major Drake, stated: "The news was received about 0800, and at 0830 I was instructed to go forward and take command of the battalion. Colonel C. T. 'Silent Buck' Lanham, the Regimental Commander, gave me only one order—on arrival I should not get into the same hole with Captain Henley. We had lost too many officers already. The first day I did little but observe and did not actually take over until about 1800." Finally, at 0945, following thirty minutes of artillery preparation and after support from the aircraft of the Ninth Tactical Air Command, the attack was underway.

Against heavy artillery, mortar, machine gun, and small arms fire, the 1st Battalion moved ahead and by noon had taken trail junction X. By 1300 hours it had pushed a defensive flank 500 yards further north. The attached light tanks failed to be of any support because of the antitank mines and dense undergrowth. As planned, the 1st Battalion then drove eastward to cut off Road A. Company B, which had been one of the assault units of the 1st Battalion, attributed its success to the scouts. As Lt. Tony Bizzaro later commented about two of his lead scouts, "Garcia and Jefferies were two of the best scouts I have ever seen. They had just plain guts and were always well forward." (Editor's note: PFC Macario Garcia was to earn the Medal of Honor on 27 Nov 1944, ten days later). Regarding Jefferies, PFC Ward of Company B relates that throughout the campaign he would never stay in his hole even during the heaviest artillery or mortar barrages but would constantly hop around firing at possible targets. "I wanna make 'em think there's a battalion here," was always his comment).

The 2nd Battalion, fighting to move abreast of the 1st, was slowed by enemy infiltration to its rear. By late afternoon, leading elements of the 1st and 2nd Battalions had attained positions from which Road A could be dominated by fire. Here they dug in for the night, covering the intervening gap with patrols.

Meanwhile, Major James C. Kemp had assumed command of the 3rd Battalion and had committed his battalion to clear out enemy groups that had infiltrated behind the 2nd Battalion. The remainder of this battalion continued to protect the southern flank of the CT. Lt. Bridgman, the new S-3, worked continuously to ensure the security of the battalion by tying in the rifle companies.

During the night, artillery fire continued to pound the Combat Team, patrols were active, and an enemy counterattack was repulsed. The plan for the next day, November 18th, was an attack to the east with the 1st and 2nd Battalions online, the 1st on the left. The 3rd Battalion was to remain in reserve, protecting the main supply route. The two assault battalions were to cross Road A, the 1st Battalion to seize Hill Y, the 2nd Battalion to seize Hill Z.

The attack moved as scheduled, progressing slowly and meticulously. The 1st Battalion, after terrific fighting, crossed Road A, advanced 500 yards, and by the middle of the afternoon had secured its objective. Throughout the campaign as the companies advanced, the leading rifle platoons were able to move ahead with the fewest casualties from artillery and mortar fire. According to

PFC Elton K. Fisher, "I'd rather lead with an assault platoon any day than wait and have to wade through all that shell fire which the support and 81mm mortar platoons always get."

The 2nd Battalion was temporarily held up by extremely accurate mortar fire. Too, it encountered an extensive anti-personnel mine field. In the meantime, Company E, which had been protecting the south flank and which had repulsed a counterattack earlier, started forward, resulting in the loss of contact and direction. It was late afternoon before contact was made by Lt. Mason's platoon. During these actions, the 2nd Battalion Commander, the Battalion S-3, and the Battalion Communications Officer were wounded. The Battalion Executive Officer immediately moved forward with a new S-3 to assume command. Within five minutes after his arrival at the CP, he was wounded, and the new S-3 killed. The Regimental S-2, Major Howard C. "Wild Man" Blazzard, then moved forward to assume command. By the time he arrived at the CP, no member of the 2nd Battalion staff remained. Assisted only by one runner, Major Blazzard pressed the attack, and by late afternoon had taken the objective Z. Here they tied in with the 1st Battalion for the night.

Because the Combat Team was fighting without protection to the flanks or rear and with a front of over 3,500 yards, an all-around defense was emphasized. A two-day supply of food and ammunition was kept moving to the assault units so that all units might be self-sustaining if cut off from the rest of the Regiment at any time. The 3rd Battalion had previously echeloned to the north, maintaining contact with the attacking units, and protecting the north flank of the Combat Team.

Throughout the day, artillery and mortar fire continued to interdict the heavily mined firebreaks, which had become a mass of mud from the constant rain. Two additional platoons of Company C, Fourth Engineers were placed in direct support of the CT, but it remained impossible to get through to the attacking units with supporting armor. Long hand-carries were necessary to supply these forward battalions and to evacuate casualties.

The advance was halted on the 19th, because to further the attack was impractical. The major reasons were: (1) A portion of Road A in the sector of CT 8 had not been opened and was still in enemy hands, as the attack of CT 8 had not developed in accordance with expectations; (2) Trail E had been so heavily mined that engineers were not yet able to clear it; even after it had been swept twice, vehicles still hit mines; (3) The hand-carry of wounded and supplies was so long by this time that further advance was an impossibility; even replacements were being used as carrying parties; (4) The bridge across the swift mountain stream near the junction of Roads A and B had been blown out. The bridge site, together with the valley, was under such heavy and continuous mortar and artillery fire, including railroad guns, that it had been impossible to bridge the stream. Carrying parties and litter squads were forced to ford the icy stream which had flooded out of its banks and was neck deep; (5) The 3rd Battalion was also experiencing difficulties; Company I, which had taken over the defensive flank to the north when the 1st Battalion turned east, had not been relieved by the 24th Cavalry Reconnaissance Squadron, although this relief had been ordered twenty-four hours earlier; Company L had been seriously disorganized by loss of key personnel and by failure of mop-up parties to return.

November 19th was, therefore, advantageously used for re-organization, re-grouping, re-supply, consolidation of positions held, and the opening of Trail E. Communications were continuously being destroyed. In order to supply the 2nd Battalion, carrying parties were organized from Service Company. The

Regimental S-2 returned to his duties when he was replaced by a new commander for the 2nd Battalion, Lt. Colonel Thomas A. Kenan.

But all of the story of war is not told in the mingled mud and blood of the front line, nor in the screaming artillery that sears and tears the flesh, nor in tales of the men who endure the fight, who curse the enemy and their own luckless plight, but who nevertheless continue to battle the most determined stand of a powerful enemy.

One of the most necessary but hardest tasks was the evacuation of the wounded to the rear. Sometimes two or three wounded men came back together, two of them able to walk supporting a third. Perhaps they had been patched up after a fashion by the forward aid man, but the casualties were too great and the wounds too many for him to be able to see all of them, and so back they came in an ever-increasing stream, some on litters having to be carried through the mud, setting their teeth at every jar, or crying out in pain as the litter bearer slipped and perhaps dropped his end of the litter. Streams had to be forded, and indeed, it seemed as if nature and God Himself had turned his face away from this embittered and tragic regiment.

At 0200 a railroad gun had fired from Duren, some five miles away, and had hit upon a dugout occupied by three officers. The dugout had a heavy roof of two layers of six-inch logs but the shell, having landed beyond the dugout, blew back in. One officer was killed outright. Another, a TD Officer, was wounded in the chest. The third, an infantry officer, had his right leg broken in a compound fracture, the shrapnel passing on through his left ankle, leaving a hole the size of an egg. Strangely enough, the pain came from the broken leg and in the dark the officer put a tourniquet on the broken right leg, not even knowing his left foot was injured. And so, he lay through the hours of the night — long, bitter, terrifying hours — while he constantly bled, growing weaker and weaker, and feeling the great grayness approaching closer and closer. Nothing could be done, for in the hell of the inferno of artillery which continued minute after minute and hour after hour, no creature could move with impunity, and it would have been sheer suicide to attempt evacuation under these conditions. Indeed, the evacuation could not be affected until eight o'clock the following morning when a litter party had to remove the two layers of logs in order to evacuate the two living officers to the aid station.

In the aid station, the battalion surgeons, working under strain, loss of sleep, and the pressure of increasing casualties, still continued to work quickly and effectively. Blood plasma, priceless life-giving fluid, was quickly rigged and administered. The wounded officer was given four bottles and now for the first time some semblance of life began to appear in his ashen cheeks, but with it, stupefying and heartbreaking pain.

They were placed in the ambulance, these wounded, two litter cases, carefully slung in racks, with the wounded sitting on the floor and on the seat along the side. Then the ambulance started down the makeshift road toward the safety of the collecting station. A man with an arm off at the shoulder tried to sit erect. The ambulance lurched as it headed for the ravine and the bridge which had been thrice blown out by enemy artillery. The driver increased his speed, for he knew there was intermittent fire on this bridge and that it was by luck and a prayer that any vehicle got across without being hit. Ambulances like any other vehicle, were fair prey for artillery. The increased speed over the rough roads, pock-marked by shell and mortar, had the effect of a medieval torture rack on the broken men within. It seemed incredible that the human body could be so desecrated and still contain the spark of life and humanity. The bridge was crossed, and the journey stretched itself into just another trial the foot soldier had long ago learned he must endure.

The collection station, set up across a German mountain stream in a German farm house, was busily working since the wounded from the entire combat team were collected here. Every wound was quickly examined, and the wounded sorted into orderly categories. The walking wounded sat in one room on the floor, or on the chairs, or simply stood, staring vacantly at one another — stood and wondered and waited.

In the next room, the litters lay on the floor so close to one another that the doctors and the aid men frequently had to step on the litter itself. Aid men quickly and efficiently appraised wounds and brought into play their first and most efficient weapon, a pair of scissors, which they carried tied to their wrists or waists by a piece of Carlisle bandage. A sergeant by the name of Stelling took a quick look at the wounded captain's feet and, grabbing his scissors, began cutting the clothing from the knee down.

The amount of clothing which the soldier wore was appalling, but he wore everything he could get his hands on in an effort to keep warm, since there were no blankets. The scissors cut through a pair of fatigues; beneath the fatigues, a pair of OD's, beneath the OD's, long underwear, and long socks; and now the sergeant saw the condition of the leg. He cut the clothing completely open to the shoe, but the foot lay twisted in an odd and somehow horrible position, and the slightest movement of the shoe or the litter caused the soldier to grit his teeth with the resulting pain. The sergeant took a razor blade and began to cut the laces of the shoe, and the pain became excruciating. It was necessary to cut the leather to remove the shoe from the broken foot, and the soldier fainted from the pain.

The sergeant had called sharply for plasma, and from a wire run across the center of the room between two windows, a T/5 had already hung a bottle, and with another stretch of bandage, had twisted the tubing and had tried to insert the needle into the veins of the forearm; but the soldier had been through too much, and from lack of blood the veins had almost collapsed. The T/5 appraised the situation and called sharply, "Captain!" A tired, hollow-eyed surgeon raised his eyes and, without a word, immediately saw the situation. He came at once and, calling for a scalpel, he slit the skin inside the elbow, exposed a vein, and expertly slipped the needle into the vein itself. Then he stood and rested his back as he watched the plasma drop by drop giving life to the almost empty veins of the captain.

After such emergency treatment as could be given had been given, the wounded were again loaded in the ambulance, while those who had died in the wretched building were carried outside and a blanket placed over their bodies — if indeed there was a blanket available. As the ambulance pulled out toward the Division Clearing Station, the medics heaved a sigh of relief and relaxation as they began to clean up the mess which resembled nothing so much as a slaughterhouse of filthy bandages and dressings, shoes, clothing, soiled and stained and bloody, and bits of human flesh and bone. As they continued their work, someone poked his head in the door and said without inflection, "There's hot coffee." The medics dropped what they were doing and picked up their canteen cups and went to enjoy this luxury, knowing full well their respite was short, and it would be only a matter of minutes until another convoy of ambulances, laden with wounded, would arrive from the death factory that was Hurtgen.

In the days through Hurtgen, the wounded were everywhere, overflowing the Fifth Field Hospital in Eupen and the Ninth Field in Yerviers, the 128th Evac outside of Eupen, the 614th and 46th at Malmady.

Incredible how these men, after having been through so much, still had but one thing on their minds—"How's the outfit?" "How are they doing?" "Is Jim still okay?" "What's happened to Bill?" A ward boy came along with a pan of hot water, the first hot water these men had had for days. Their faces were still caked with mud, their bodies filthy, they were unshaven, their hair was matted and uncombed, their teeth caked, and their mouths dry, and then suddenly there was hot water and soap and gentle hands to shave them and someone to help them. This was a world they had scarcely remembered, with its orderliness, its friendliness, and special care for them, its attempt to ease their pain and do what it could for them; and yet in the midst of this bright new world they were suddenly very much afraid and very much alone, because they had left not only parts of their body but a part of their soul and heart with the Twenty-Second Infantry Regiment. And just before they dropped off to sleep there flashed through the recesses of their minds the Boll Weevil with its blue parapet against a white field and below it the words which they somehow seemed to see confused with splintered trees and broken men, "Deeds, Not Words."

During the grueling days of the battle and occasioned not alone by the fighting itself, but aggravated by the abominable weather, were the cases of combat fatigue and exhaustion. Known as shell shock in the last war, the condition had been dressed up in this one in an attempt to alibi or to kid the soldier into something he himself should believe, but which he seldom did.

Throughout the first days of the battle, a sergeant in a Heavy

Weapons Company had been used as a mortar observer. Again and again, he had gone forward of the lines, dug into a position, and there observed and adjusted mortar fire on enemy positions. On the night of 23 November, the sergeant was again ordered forward, and he went, taking his radio operator with him. When they arrived at their position for observation it was just turning dusk. Suddenly a shell came in, landing close to them, and killing the radio operator. The sergeant strapped the radio on his own back and reached for his shovel so he could dig in for protection against the constant artillery when a machine gun bullet clipped the shovel, cutting the handle completely from the spade. For thirteen weary hours the sergeant lay on top of the ground in the midst of one of the worst artillery and mortar barrages the battalion could remember. He himself counted thirteen hundred rounds in his immediate vicinity, and then his mind went blank, and he quit counting.

At nine o'clock the following morning, the radio again came through and he was ordered back for relief. He picked up the radio and stumbled toward the rear, eyes staring straight ahead, mind vacant, lips slightly drooling. He hit the ground when shells came near but got up and stolidly walked on back when the concussion was over. When he reached his company, his first sergeant took a quick look at him and said, "Sergeant, go to the aid station." Without uttering a word, the sergeant stumbled on back, dropping the radio as he went. When he got to the aid station, he simply sat down and waited. A medic came by and said, "What's the matter with you, soldier?" but he did not answer, only stared. The medic snapped his fingers a couple of times in front of the staring vacant eyes, and then wrote on the tag "Exhaustion." He had to wait for the first three ambulances since they were full of wounded who needed immediate attention, but into the fourth he climbed obediently as instructed and sat on the floor.

Through the usual chain of evacuation, he came at last to the 622nd Exhaustion Center, stationed outside of Eupen on the Vervier Road.

The standard treatment upon the arrival of an exhaustion or combat fatigue case was to give the man narco-therapy. This consisted of enough sodium amytal to keep the man sleeping constantly for three days. He awoke only long enough to go to the latrine and in most cases, he had to be helped even for this. Frequently he was given saline intravenous injections in lieu of feeding. During this time his subconscious condition was horrible to behold, and he lived over and over again his most terrifying events of combat.

When he awoke, he was hollow-eyed, cheeks drawn—the shock upon both his physical and nervous system had been tremendous. He was taken to the showers where he enjoyed a hot, invigorating shower, perhaps for the first time in weeks. Subsequently it was a matter of care, perhaps of psychoanalysis, which at best was but fleeting, due to the number of men and the scarcity of psychiatrists. An attempt was made to rehabilitate him for use in the front lines—which never worked. Having once found that he could be evacuated with combat fatigue, the soldier instinctively knew that the medical tag reading "Combat Fatigue" represented safety for him and surcease from danger. Frequently some soldiers made the circuit from the rest center to the front lines three or even four times. It was another one of the hells of war. The psychiatrists had no choice for there was no other place to send the man unless he was physically disqualified or mentally unstable. If he were simply and truly exhausted, he went back to the front lines.

It was pitiful to observe the genuine fatigue cases in the tent wards of the rest center. It was time for chow and hot chow was being served. During the serving, an artillery observation plane flew overhead. The effect was amazing and pitiful. One soldier poured a cup of hot coffee over his head. Another turned his mess kit of C rations over in his lap. Still a third made an effort to dive in the latrine and had to be forcibly restrained. Some of the others crawled underneath canvas cots for instinctive protection. It was still true that the psychology of sound was stronger than the psychology of the sight or of the mind, for this was a place and a time where the instinct of self-preservation was strongest.

The wards with shattered and missing legs and arms were bad, but the hospitals with vacant and missing minds were worse.

Up front in the half light of the forest, in the frozen slush and deep silence punctuated by the staccato of the German burp gun fire and, answered by our M1 Rifle and light machine gun fire, the fight went on. Exhausted men swore softly and died or cried and collapsed as was their want, but the Twenty-Second Infantry Combat Team of the Fourth U. S. Infantry Division fought and slugged and pushed and dragged its way eastward yard by yard, 500 yards a day, 1,000 yards a day, 850 yards a day, 85 casualties a day, 143 casualties a day, leaving a trail of broken men, American and German side by side, to mark its progress.

* * * * *

The After Action Report of CT 22 for the month of November 1944 describes the events of the first several days in Hurtgen Forest as follows:

November 16. At 0100 hours Combat Team 22 was notified that

D-day was 16 November and later that H-hour was 1245 hours. Several heavy artillery concentrations fell in the area during the night 15-16 November and some casualties were sustained. Shortly after daylight, a German patrol engaged a Twenty-Second Infantry wire team. This was the first actual contact with the enemy in the sector.

Combat Team 22 planned to attack (initially) in a column of battalions in the order 2nd, 1st, and 3rd. The 44th Field Artillery Battalion (reinforced by the fires of the 20th Field Artillery Battalion) was in direct support. Attached to Combat Team 22 were medium tanks of Company C, 70th Tank Battalion, light tanks of one platoon of Company D, 70th Tank Battalion, 4.2 chemical mortars of Company C, 803rd Tank Destroyer (self-propelled) Battalion, one platoon Company C, 4th Engineer (c) Battalion and Company C, 4th Medical Battalion. As the roughness of the terrain and the denseness of the forest precluded their use, the armor and tank destroyers initially were held in reserve.

In the four daylight hours of its first day of attack, Combat Team 22 gained approximately 1,500 yards at a cost of 46 enlisted men and 9 officer casualties. Enemy artillery and mortar fire increased in intensity.

November 17. The scheme of maneuver for the attack on 17 November called for the 1st Battalion to continue its advance along Trail E to seize the hill approximately 1,000 yards to the north and trail junction X on its forward slope. The battalion was then to swing east and attack abreast of the 2nd Battalion to seize the high ground approximately 500 yards to the front which dominated Road A.

The attack, scheduled to jump off at 0830 hours, was delayed until 0945 hours by extremely heavy enemy artillery and mortar concentrations throughout the Combat Team sector, the death of the 1st Battalion Commander, the wounding of the 3rd Battalion Commander, and extensive mine fields. However, with a thirty-minute artillery preparation, and with pursuit aircraft of the Ninth Tactical Air Command in support, the 1st Battalion attacked at 0945 hours.

In the second day's operations, Combat Team 22 advanced approximately 1,000 yards to dominate Road A—the main north-south road through the Hurtgen Forest. The Combat Team sustained 104 enlisted men and 2 officer casualties during the day; received 148 enlisted men replacements; and captured 44 enemy. Of importance in the enemy's defenses were his extremely heavy artillery and mortar concentrations on all trails and firebreaks which could be used as supply routes. Engineers were seriously hampered by this heavy fire in the mine sweeping and road maintenance necessary for supply and evacuation and for supporting armor.

November 18. During the night artillery continued to pound the Combat Team, and the enemy attempted to infiltrate 2nd Battalion positions. Our patrols probed enemy lines during the night. A small-scale counterattack against the 2nd Battalion at 0658 hours was quickly repulsed by artillery and small arms fire.

The plan for 18 November was an attack to the east at 0830 hours by the 1st and 2nd Battalions abreast, 1st Battalion on the left, with the 3rd Battalion in reserve protecting the main supply route. The 1st and 2nd Battalions were to cross Road A and seize Hills Y and Z respectively.

The attack jumped off as scheduled and advanced slowly, the 2nd Battalion against heavy machine gun and small arms fire, and the 1st Battalion against heavy artillery and mortar fire.

Fighting hard, the 1st Battalion crossed Road A at 1013 hours and from there advanced an additional 500 yards, seizing objective Y by 1430 hours. The 2nd Battalion was held up by heavy and extremely accurate mortar fire. In addition, it encountered an extensive anti-personnel mine field which was found to be both wide and deep. Considerable time was spent in finding a way around these mines. In the meantime, Company F, which had been protecting the right (south) flank and which had beaten off the enemy counterattack early that morning, started to move forward and in so doing lost both contact and direction. The Company was not located until late in the afternoon. Finally, the battalion commander, the battalion S-3, and the battalion communication officer were all hit and evacuated.

The battalion executive officer moved forward with a new S-3 to assume command. Within five minutes after reaching his command post, the new battalion commander was hit, and the new S-3 was killed. The Regimental S-2 was then ordered to move forward and assume command of the battalion. By the time he arrived at the command post, no member of the battalion staff remained. Assisted only by one runner, the new battalion commander jumped off the attack at 1430 hours, and by 1650 hours had reached objective Z. At this point he was joined by a former officer of the regiment who had just returned from the hospital, and who was to act as his executive. The 1st and 2nd Battalions tied in and secured for the night on their objectives."

The Combat Team gained 1,000 yards of forest on 18 November at a cost of 150 enlisted men and 13 officer casualties; 43 prisoners were captured.

EXTRACTS FROM THE UNIT JOURNAL FOR 18 NOVEMBER GIVE THE FOLLOWING PICTURE:

DATE: 18 November 1944
WEATHER: Fair
PLACE: 986342

Time | To | From

0820 | S3 | Red 6
We are receiving heavy mortar fire.

0840 | S3 | White
Moved out on time

0855 | S3 | Red
Red jumped off at 0830.

0927 | S6 | S3
Red and White moving slowly. One platoon is up to the road—no one across receiving mortar, arty, and SA fire (this report has been sent to G3).

0935 | Red | S6
You'll have to move faster to reach edge of woods before dark. White informed thusly. Red said they are 100 yards from road moving slowly.

0950 | S3 | White
Company G pinned down on right by MG fire.

1050 | S3 | Red 5
We are 150 yards short at (027387), about 400 yards beyond roadblock. Company B has been relieved. My communications officer has been hit – have Captain Hickey send me another.

1055 | S3 | White
We are running into a lot of arty and SA fire, haven't crossed road yet.

1100 | Red 5 | S3
Are Engineers clearing the road? Yes, but it is being shelled heavily. Tank knocked out at (009379), S3 tells Red mediums are available at roadblock #2.

1115 | S2 | G2
Holliday (28th Division) reports enemy tank at (008378). shelling this coordinate heavily. S2 reports they are our tanks. Tell Holliday to quit shelling.

1145 | S3 | White
(Via radio) Charlie 6 is now in command White Bn.

1152 | S2 | White
Seven PWs on their way down.

1205 | S3 | White
We are getting plenty of arty here. Colonel Walker has been hit.

1345 | Blue | S3
First Bn. Is moving 500 yards across road. Follow and protect his flank. Second Bn. held up. * * * * * Blue 6 with I and K Company and Engineers to clear Road also helping on bridge. S3 says he will try to get

more Engineers to relieve men on bridge so they can work on road. Blue 6 is at016385. S3 tell him to patrol road to boundary until more engineers are there. Maintain contact at 018382 with Red and White.

1600 | S3 | Red
We have word from White that they have advanced 500 yards and they are still moving. Leading elements at 027384.

Incessant enemy artillery and mortar concentrations continued throughout the night of **19 November,** but the Regiment again attacked at 0850 hours the morning of the 20th. The plan was: First and Second Battalions abreast (Second on the right) moving along axis of Road B to seize the dominating terrain 600 yards to the east. The Third Battalion, in reserve, was to open Road A as an MSR, clearing it north to the CT boundary.

The attack by the Second Battalion met terrific enemy artillery, mortar, and small arms fire, and it was later determined that the Germans had launched an attack at the same time. The Second Battalion, advancing slowly, reached the objective before noon. The First Battalion, against light resistance, also took its objective before noon and at once blocked Road B with mines and then covered these blocks with small bazooka teams. An enemy counterattack from the north was hurled against the First Battalion only twenty minutes after they had taken their objective. The battalion waited until the enemy was within close range before opening fire. This counterattack was rebuffed, and the enemy suffered a very large number of killed and wounded. Immediately after this was repulsed, another even larger counterattacking force known to include six tanks or self-propelled guns struck the Second Battalion to the southeast. Company L quickly moved to re-enforce the Second. The attack was stopped, and the situation cleared by 1300 hours. Company L remained attached to the Second Battalion to cover their right flank.

The First and Second Battalions were well established on their objectives by 1430 hours, **20 November.** However, as before, major difficulties still confronted the CT: (1) Forward units and all possible supply routes were constantly subjected to heavy artillery and mortar fire. (2) The one available motor route, Trail E, was an axle-deep, boggy marsh which had not yet been cleared. Along this trail, into which vehicles were canalized by the thick woods, the enemy had buried mines three deep so that they would not be discovered nor explode until deep ruts had been cut in the mud by traffic. Box-mines were found along the shoulders and ditches bordering the trail. Anti-lifting devices were attached to the great majority of these mines and thus necessitated destroying the mines in place, and additional work to fill the resulting craters. As a result of these conditions, a hand-carry of more than 1500 yards was necessary to supply forward elements and evacuate casualties, and armor and antitank weapons were unable to get forward.

At the Regimental CP, an attack was received and the Hdq. Commandant was killed. Company K counterattacked that afternoon to destroy a by-passed enemy strongpoint. The strongpoint had not fallen by dark, and resumption of the attack was planned for the next day.

The next morning, **21 November,** the Division Commander directed CT 8 to attack south to open that portion of Road A in its sector, and in addition to clear out the northern portion of its sector between Road A and Trail E. A large patrol moved without difficulty and without hostile contact to a previously designated spot, and there awaited contact with CT 8. By late afternoon, contact had not materialized, and the Division Commander directed CT 22 to organize a reinforced company and attack beyond its boundary until contact was established. Company I, plus tanks and tank destroyers, attacked north along Road A, swept the area, and contacted Company L of CT 8 at 1655 hours without encountering the enemy.

Earlier the same day, Company K, supported by a tank and a tank destroyer, had resumed the assault on the enemy strongpoint near the Command Post of the Combat Team. The enemy was crushed, and 20 prisoners taken.

Casualties were still extremely heavy, and the Germans had brought railroad guns of huge caliber into the battle. Replacements were arriving daily to take the place of the wounded or killed.

The CT resumed the advance on the **22nd of November.** Just prior to 0830 hours, following a night of moderately heavy artillery and mortar fire, the First Battalion feinted a frontal attack, thereby diverting the enemy. The Third Battalion, having previously reassembled, swung north from its reserve position, and attacked around the left flank of the First Battalion an hour later and drove southeast. The envelopment was highly successful. Although contact with attacking echelons was lost when key personnel became casualties, the advances of the Third Battalion, commanded by Major James C. Kemp, against all types of fire carried some 1200 yards eastward to within sight of Grosshau. The battalion now dominated and cut by direct fire the main road junction 700 yards west of Grosshau, and so consolidated its positions that it cut the roads leading to this junction from the west and northwest.

The Second Battalion that day was to advance south of Road B toward Grosshau, the division objective. The Battalion, repulsing an enemy attack on its north flank, jumped off at 0930 hours. Immediately, fiery resistance was encountered; but despite severe casualties, it pushed across Road C and gained nearly 1000 yards. One hundred Second Battalion replacements received during the afternoon had to be used to cover the south flank of the Battalion, since it had become more exposed as the attack progressed. The First Battalion, now in reserve, dispatched one company south to protect the right flank of the Second Battalion.

THE UNIT JOURNAL REPORTED THE ABOVE ACTION AS FOLLOWS:

DATE: November 22, 1944
WEATHER: Rain
PLACE: 006374

(Note: Red is 1st Battalion, White is 2nd Battalion, Blue is 3rd Battalion — S6 is regiment commander, G6 is division commander, S3 is regiment's operations officer, S2 is intelligence of=ficer, 5 is executive officer.

Time | To | From

1202 | Blue | S6
What is stopping your progress? If you do not make swing, we will wind up in a bad position. You must make speed in the last few daylight hours.

1236 | Red | S3
Have you heard anything from White? A: Yes, they are pinned down by MG fire.

1305 | S6 | White
We have suffered heavy losses, but we are progressing and evacuating our wounded.

1301 | S6 | S2
The Engineers refuse to sweep the main road to allow the tanks to get up to White. S6 orders Engineers to do this as men are dying for the need of those tanks. TDs are getting up to White.

1505 | G6 | S6
Blue is on its objective. Grosshau is filled with Krauts and we are firing a lot of artillery in town. White has suffered heavy losses. G6 warns that we are to be set for reactions.

Operations on the 23rd and 24th were limited to consolidation and readjustment. Road B was to be cleared to the road junction dominated by the Third Battalion; armored support and antitank weapons were to be moved to forward elements; and four limited objectives were to be taken, four key trail junctions just ahead of the front lines.

The Second Battalion consolidated its position during the 23rd and gained control, by fire, of the two trail junctions given as objectives. A small task force of the First Battalion, including a platoon of tanks and a mine-sweeping detachment of Engineers, moved east along Road B and despite artillery and mortar fire were able to clear it by 1330 hours. The Third Battalion had direct fire from Grosshau on its positions the entire day but lost no ground.

An early attack was planned to take place on the 25th of November. The 2nd and 3rd Battalions were to be the assault units. There was to be no artillery in an effort to effect surprise. The scheme was: The 3rd Battalion would envelope Grosshau to the north and take the town from that direction; the 2nd Battalion would move to the eastern edge of the woods from which direct fire could be placed on both Grosshau and Kleinhau; and the 1st Battalion, in reserve, would move east, following the advance of the 3rd Battalion with one company astride Road 'B'. Additional replacements, 184 enlisted men and 25 officers, were received the night of 24 November and immediately sent to the forward units.

The attack was underway shortly after daylight, 25 November, and the 3rd Battalion rapidly pushed northeast 800 yards to the wooded area north of Grosshau. As Capt. Roche later said, "On that drive we went through the Krauts like a dose of salts, at least we got a good start." After a quick reorganization, the Battalion prepared to assault the town of Grosshau. (All the information the Regiment had from Division G-2 indicated that the town was lightly held.) The Battalion, starting across the open ground north of Grosshau, met immediate and powerful enemy reaction. German self-propelled guns, mortars, and small arms stopped the attack and disabled four tanks and two tank destroyers. All attempts to take the town suffered tremendously, and before dusk the Battalion retired and dug in for the night.

THE UNIT JOURNAL REPORTED

DATE: 25 November, 1944
WEATHER: Rain
PLACE: 006374

Time | To | From

1317 | Red | S6
Has Red 6 a reading with Blue? Want him to have the real score. Have a sad story from C. O. of tanks, don't know how much of it to credit. Tell Major Goforth to move the rest of their Company to clear the road and then we can use the rest of the battalion to aid Blue.

1542 | S6 | White
Capt. Eggleston fatally wounded. Capt. Daniels will act as Ex.

1545 | S6 | Red 6
All tank officers are either wounded or knocked out by concussion. The TDs have been suffering also. Blue is in a pretty bad way. Casualties are running high. S6 says the road must be kept open between Red and Blue. Watch your left flank. Try to keep physical contact with Blue. You will be sure of a counterattack in the morning. Red reports the men are pretty well worn out. S6 says that G6 and he both realize this but there is nothing in the way of relief in sight. The General has appealed to higher headquarters but to no avail. Our condition is better than either that of the 8th or 12th.

1610 | Pro. | Corres.
Mr. Hemingway said that when Capt. Stevenson calls, have the call put

through to S6 phone as I will "be with the 6 for some time." (Reference to War Correspondent Ernest Hemingway.)

Tanks were completely unable to reach the 2nd Battalion which had reached the edge of the woods west of Kleinhau, because of mud and poor trails. That unit, without armored support and against bitter resistance, had pushed the necessary 800 yards to its objective. Late in the afternoon, tanks arrived at the forward positions. The 1st Battalion moved eastward to the position held the previous night by the 3rd Battalion.

No major attack was planned for the 26th, but there were objectives to be taken. During the day the 81mm. mortars and artillery pounded the well defended town of Grosshau. Company "C", astride Road B, attacked east to clear the woods west of Grosshau. Due to a narrow corridor into the town from the west, the use of a larger force was precluded. The company met stubborn resistance, including direct artillery fire, but by late afternoon had taken the objective. The intended AT support did not reach the company. At dusk a German counterattack supported by self-propelled guns hit the company and drove it back to the positions held the night prior. The company sustained heavy casualties and by the close of day had an effective strength of only twenty officers and enlisted men.

In the past week of bitter day-by-day, inch-by-inch fighting, there were unarmed soldiers daily sacrificing their lives in order that others might live. These were the aid men, the litter-bearers, and the chaplains. These aid men, working throughout the Combat Team, on the front lines and in the aid stations, never once hesitated to do a job assigned them. Of course, they were scared; it was impossible not to be afraid; yet these aid men would crawl out into the face of certain death in order to bring help to the wounded. Their job was a detestable one; for they had no weapon to fire, and they were forced to look at torn limbs, chopped flesh, and the dead men. They knew what one shell fragment could do to the human body. No, their job was not a glorious one; it presented no spectacular results, no ground gained, no enemy killed, but only the hideous, never-ceasing task of repairing torn human bodies. How long could a man's brain endure the hellish sounds of dying men? Only aid men and litter-bearers knew the answer. They had rather die themselves than to give up. It was remarkable to note the small number of cases of combat exhaustion that resulted from the strained bodies and minds. The chaplains worked feverishly to bring aid to the wounded, the fighting men, and the raw replacements. Their jobs were not on the front lines, but that is where they were to be found a great amount of the time. Chaplain Boice and Chaplain Hogg, each working with his own battalion, helped administer blood plasma and to instill new courage into those affected. The chaplains were not surrounded with any heavenly cloak to protect their bodies from artillery, yet they went to the front in the face of death to pray with the men and to bring new fighting strength to the unit. What glory have they? None, except the answering prayer within their own hearts that they were accomplishing the mission for which they were intended. Never let it be forgotten that an infantry rifleman stands not alone in battle; he is surrounded by men without weapons who are ready to help him in time of need. The chaplains and the medics of the 22nd were heroic men.

During the night of the 26th, a patrol from the Third Battalion investigated Grosshau, returning with the information that there was considerable German activity within the town entrenching and the sound of tracked vehicles was heard.

At 0900 hours the 27th, Company B, under direct fire, attacked with 105 men against the Germans firmly entrenched on a dominating hill. Lt. Murray, with the seventeen men of the First Platoon, was leading. For two hours the platoon tried to advance and succeeded only in crawling up five or six yards closer. Lt. Murray was still in the lead but fortunate enough to have shelter in a large shell crater. The rest of the platoon behind him was annihilated; all seventeen were either killed or wounded. The Third Platoon tried to advance but were also cut to pieces. The Second Platoon then tried to slip across the field a squad at a time in a skirmish line. Ten men got across the clearing and within twenty yards of the German-held woods, when a machine gun pinned them down helplessly. Pfc. Charles Edwards was the first who endeavored to destroy the gun. Edwards, a former member of the Fourth Engineer Battalion, had requested that he be sent to Company B; this was granted near Brandscheid. Edwards crept up toward the gun but had not advanced more than five yards when a quick burst from the machine gun killed him.

S/Sgt. Thomas Dyess tried to worm his way around through the woods to flank the gun from the right, but he too was wounded in the attempt. Pfc. Macario Garcia, acting squad leader in the support platoon, then went into the woods. Several grenades were heard exploding and Garcia emerged from the woods announcing, "God damn, I killed three Germans and knocked out the machine gun." No sooner had this been said when another machine gun opened fire and Garcia, though wounded, re-entered the woods, completely annihilated the machine gun crew of three and took four prisoners without assistance. As he came out this time he said, "That's all of the bastards here." This one man's efforts and success enabled the remainder of the company to advance on into the woods. (Note: Garcia was later awarded the Congressional Medal of Honor for this act.) By this time, Company E arrived on the double to assist Company B, having been detached from the Second Battalion to the south.

Following are excerpts from notes kept by the Commander of Company E, Captain Faulkner: "On Monday the 27th, I was called to the Second Battalion CP (only a hole for three among the small pine woods back across the firebreak) and received an order from Colonel Kenan to take Company E at once 600 yards to the north, to hit the Kraut on the flank, and assist Company B which was bogged down, then to return to our own area. We left our weapons platoon with the Executive Lt. Mason to hold our own position and started out through the pines toward the north—79 enlisted men and 5 officers.

"After moving only several hundred yards we received a radio order from White 6 (Second Battalion) that Company E was no longer under his control but now attached to First Battalion and to rush to the aid of Company B fast—they were hurting.

"We moved out at the double and down into a stream draw. I thought: 'This is a natural spot for a mortar concentration,' and either the Kraut had mental telepathy or good observation, for immediately a heavy mortar concentration swished in through the trees, smack dab on top of the command group. My runner, carrier, was injured at my side and Peters was killed near me.

"We kept going and came in on the run to a picture of real carnage—arms, equipment, dead and wounded, Jerrys and GIs strewn all through the woods. Blasted trees, gaping shell holes, and the acrid smell and smoke of small arms and mortar fire completed the terrible scene. It looked just like Hell!

"We found the acting Company Commander, Lt. Pizarro, in a hole up the hill. Boy, was he glad to see us. We moved on to make a quick recon under sniper fire. With Lt. Lloyd's excellent help, and under sniper and mortar fire, our men were strung around the hill and we picked up a few Krauts on the way. We could see Grosshau only several hundred yards to the east beyond the edge of our woods.

"We dug in at once. Company B remains—2 lieutenants and 15 to 20 enlisted men pulled out after dark, a couple of TDs moved up to help us hold and we were on our own. A blanket of smoke came down over our position. On checking, we determined it apparently was not ours, so the Krauts must be getting ready to counterattack. This was no picnic.

"On the 28th, we had 88 and mortar fire all day. The rest of the woods fell down over our heads. We had several casualties, and we dug deeper and deeper. We also dug new holes for the 110 replacements coming to us that night. Only about 90 arrived with Lt. Mason, the rest having become casualties while moving up through a barrage. On orders from Red 3, I sent a 12 man patrol under Lt. Larson into Grosshau at 2005. At 2010, they breezed back like a bunch of big-tailed birds, followed by burp gun fire. Town full of Jerrys."

In conjunction with the action of Pfc. Garcia the afternoon of the 27th, there was another similar display of individual heroism. Lt. William Jordan, from Regimental Antitank Company, volunteered to take his 57mm gun and crew down an open, exposed road in an attempt to destroy the German tanks firing so accurately on Company B. He told Captain Henley that he would take his gun and half-track down himself, allowing his crew to proceed on foot. He set up his 57mm gun at the north edge of the woods in the front line of Company B and was followed by four tanks and two tank-destroyers. Captain Henley later declared, "These are two of the most heroic acts which occurred during the entire operation."

By 1440 hours on the 27th, Companies B and E had secured the woods on the western edge of Grosshau.

THE UNIT JOURNAL REPORTED THE ABOVE ACTION AS FOLLOWS:

DATE: 27 November 1944
WEATHER: Fair, Cold
PLACE: 025383

Time | To | From

1105 | S6 | G6
S6 said, "We are going slow. Have met a lot of resistance, hard fighting going on. S6 said, "Germans are throwing everything."

1120 | White 6 | S3
S3 told White to get ready to attack toward Red. North flank, said it might ease pressure on them.

1125 | Red 6 | S3
S3 asked if 'E' Company could help him (Red 6) by attacking north to flank. Red 6 said, "yes." S3 asked situation. Red 6 said not too good. Red 6 said White should attack patch of woods west of town, but should not fire - - only when they see Krauts.

1135 | White 6 | S6
S3 told White 6 that 'E' Company would have to go in with determination to help Red Battalion because that point is causing a stoppage.

1251 | S3 | White
White said 'E' Company was on the way.

1308 | Red 5 | S3
S3 told Red that 'E' Company is on the way and have they seen any-thing of them. Red said not yet and said they were having awful hard time. Red said White may be a big help.

1335 | G6 | S6
S6 said General Collins (Corps CO.) just left here and is on the way up to see G6. S6 said he had a talk with him and he agreed on taking the ridge. S6 said General Collins asked how many more days the 22nd Inf. Could stay in action. S6 said he told him about 3 days if they are very rough days. S6 told G6 that 'E' Company had been attacking since 1230 hours.

1405 | White 6 S6
S6 said 'B' Company is shot up and White Company 'E' is to attack right straight through to the road and to go in with determination and break this thing. S6 said this is different from first plan but will have to be done because 'B' Company is in bad shape.

1440 | S6 | Red 5
Red said they were now on objective with 'E' Company. Also they are digging in now. S6 said 'E' Company is now under control of Red.

1450 | S3 | S6
S6 said to get following message to G6 who was with General Collins. "Our 'B' and 'E' Companies are at edge of woods on the point of town. 'B' said only 15 men left digging in."

1648 | S6 | S3

Received radio message from tank company that Krauts are laying down a heavy smoke screen in White area, probably going to get another counter-Attack. Be sure to inform Canean and also Cactus (S6).

1656 | Red 5 | S6

Red 5 said he would like TD to have tanks and TDs on up to curve in road. Tanks and TDs refused to move up there because they were fixed upon there yesterday. Tanks are 600 or 700 yards below us. S6 said to order them up there. They are under Red 6's command.

1703 | S3 | Liaison

The 22nd will continue attack tomorrow with the capture of Grosshau and the high ground northeast thereof (orders of Cactus 6). |

1710 | S6 | Mag TDs

Tanks won't move up. We (TD's) will follow tanks if they move up. Tanks afraid direct fire weapons. S6 said there are no direct fire weapons in there. Red 5 then got on phone and said he has a boy in AT Company that said he will take a half track down road with AT and if he gets through rest of AT guns will follow.

1715 | S3 | White 2

I went out and turned tanks around and sent them back up to front lines. We are getting very heavy artillery barrage n area now.

1741 | AT Co. | S3

Temperature report. Max 50, Min 33, light rain will start at 0800, possibly turning to snow by 1000. Intermittent rain throughout period.

1855 | G6 | S6

Gave G6 the information of tonight's activity. We are waiting for time on tomorrow's attack. Blue is battered pretty badly. G6 said he could move Cargo Blue over to take White's area. (S6) believes Blue's effectiveness will cease after an hour and a half of getting shelled too heavily. The town isn't anything. If we get the town, we will get beaten to death. Got to have high ground. Our south flank is too wide open. G6 said the people on the will go early in the morning and it will be more than just infantry. Take the town of Grosshau first and then go to other objectives. S6 said he isn't interested in town, only to get main road open. G6 said that the corps arty will be available to him. S6 said he would like to wait until Cargo took Hill 90 before jumping off. We could give them supporting fire. |

2125 | G6 | S6

Told General Barton of patrol getting shot at in Grosshau. We are firing mortars and artillery in there now. General said get ready for a counterattack tomorrow.

On 28 November, CT 12 took over a sector north of CT 22, thereby reducing the frontage to 1200 yards. A task force from the Third Battalion, coordinated with CT 12, attacked northeast early that morning and by 1350 hours had secured its objective, Hill 90, north of Grosshau. The balance of the CT, under constant artillery and mortar fire, consolidated its positions.

On 28 November the Combat Team sustained 113 enlisted men and 4 officer casualties; received 9 enlisted men replacements; and captured 23 Germans.

November 29. The scheme of maneuver on 29 November called for an attack by the Third Battalion to bypass Grosshau to the north through the edge of the woods and seize the dominating high ground northeast of the town. This hill on the Hurtgen Ridge overlooked the eastern approaches to town. The First Battalion was to follow the Third Battalion and protect its flanks. The Second Battalion was to contain Grosshau and protect the Combat Team's right flank south of the town.

Ordered to attack at 1100 hours, the Third Battalion was delayed by heavy enemy artillery fire and jumped off at 1200 hours. The battalion pushed north-east against continued determined resistance, and at 1620 hours was 700 yards north of its objective and ready to swing south. Resupply was difficult and heavy fire was received from Hill 92 (to the northwest) which was not completely controlled by Combat Team 12. Despite the approaching darkness, the battalion was ordered to continue the attack; and at 1830 hours reached its objective—the hill northeast of Grosshau.

The First Battalion moved to the east, and by 1445 hours had occupied the original positions of the Third Battalion. The First Battalion remained there prepared to assist in the capture of Grosshau.

In the late morning, the Chief of Staff of Division called the Combat Team Commander, and in the name of the Division Commander, directed that Grosshau be taken that day regardless of cost. This superseded the previously approved plan of encircling Grosshau and then calling for its surrender under a flag of truce. In accordance with the new order, the Second Battalion was given the mission of assaulting Grosshau from the west. At 1250 hours the Second Battalion assaulted Grosshau with Companies E and F. Simultaneously with this attack, tanks located on the south (right) flank of the Second Battalion were to assault the town from the south.

Immediately upon jumping off, Company F (on the right) received a counterattack which it repulsed. Company E advanced slowly against heavy German small arms fire. Fighting was intense and the attack had carried only 75 yards beyond the first house in the town by 1635 hours. In the meantime, the tank attack from the south had run into heavy minefields and two tanks were lost. The attack was further canalized by a bog, but the tanks continued probing to find a route into the town. The battalion was ordered to continue the attack regardless of nightfall, and the infantry-tank attack from the west progressed slowly. The tank attack from the south finally penetrated the mines and made contact in the west part of town with the force attacking from the west.

By 1915 hours, the town had fallen. Immediately, mine-sweeping detachments moved forward to clear the roads of the many mines encountered. Later examination disclosed that, in reality, the town was fortified. Buildings were found with reinforced basements, complete to the extent of having steel doors and firing slits. More than 100 prisoners were taken from the town.

During the day, Company C remained in position securing the right flank of the Combat Team southwest of Grosshau.

Patrols from the Second Battalion (in Grosshau) and the Third Battalion (on the hill to the northeast of the town) established contact at 2304 hours. Antitank weapons were rushed forward to support the Third Battalion and to further reinforce the defenses of Grosshau. Supplies also moved forward as the road was opened. Tanks and tank destroyers were disposed to hold the town and cover all approaches.

Some additional small arms fire was received from basements of the town. Seventeen additional prisoners were taken from the buildings of Grosshau during the night, and 18 more Germans were captured there the next morning.

At 1630 hours, the 46th Armored Battalion was attached to Combat Team 22 and prepared to attack in coordination with the rifle battalions on **30 November.**

During the day, the Combat Team sustained 158 enlisted men and 4 officer casualties; received 78 enlisted men replacements; and captured 155 Germans.

The attack on Grosshau was reported in Captain Faulkner's notes as follows:

"**November 29** was bright, clear, and cold and Company E was still held. Received radio order at 1120 to attack Grosshau at 1245 with four tanks coming up. All a rush, but we moved over the LD at 1245. First platoon pinned down at once about 50 yards out by cross MG fire from village 150 yards east. Send second platoon around right of woods. They were immediately pinned down by fire. More casualties. Third platoon stopped cold at road on our left. Few dead and wounded. No 536 radio communication with platoons so used runners. Too slow. No 610 arty radio, out! Tried mortars with Lt. Fitzgerald observing—no go. Then artillery with Lt. Larsen as FO up at point of woods passing corrections along through men in shell holes. Can't slip our arty fire into edge of town where it's needed. Can't get observation.

"Hurry! Hurry! Here comes Jerry artillery, 105 and some 88 direct fire on our position. I'm in a shallow ditch with Sussman, our 300 radio man.

"First German concentration fires about 20 to 30 rounds. A minutes lull—then—swish, boom, Sussman is blown over on his back. I'm covered with dirt and can't hear. There's the hole two feet away where the Kraut shell landed. Too close. We've got to get a move on—quick.

"Here comes our tanks. Lead tank's bed rolls are on fire. Tank Commander cuts them off. First tank reports it is filled with fumes and goes to rear. Others wait.

"Time drags on and my mind twists and squirms to find a solution and a way in or to get my men out from under that murderous MG fire. Lots of casualties pouring back or lying out in the open field.

"I report our situation to Battalion on the 300 radio. A pause, then their reply: 'You'll have to do it—or else—understand? Watch for Big Boys from South—Roger.'

"Here comes our four tanks again at last. I'm shouting the order to move in with them. The gun jams on the lead tank. Lt. Yoeman, the tank commander, stops and cooly steps out of the turret with his ramrod to clear it. More Jerry mortar fire drops in on us. We can't stay in this exposed position a minute longer. I move on in the open, wave the second tank on and got our men moving on the double. Sgt. Ivy jumps to the top of the lead tank and spots strong point to the left of town. Wave on the bulldozer tank and two TDs to the left of town with first platoon and Lt. Railton following. Two tanks down the center street with second platoon and Lt. Larsen on left of street, and third platoon less one squad under Sgt. Richardson on right of street.

"Sent one squad to right of town to cover the backyards on our right flank. Kept 4th platoon now acting mostly as riflemen in reserve.

"We are moving and firing as we plug along. There comes the 'Big Boys' from the south across the open fields. Twelve of them—no, only 11 now, as one just stopped by a mine. I can see the crew jumping out.

"Well, that was it. The tanks blasted a house at a time with

HE and tracer. Their MGs went like a bat out of hell. Regular movie stuff—noise, fire, smoke, movement, confusion.

"Our boys threw grenades in the house windows and went in after the Krauts. Some 20 Jerries came pouring out of the first house. It was now 1630. We did the town that way and I followed the lead tank. I still owe the tanker, Sgt. Hogan, a pint of bourbon for his work which I promised in the excitement. (P. S. He was wounded a day later at GEY). We hit the east end of town at 1830 and reported it to White 6 on the radio. 'Nice work' was the reply.

"We set up our defense on the east edge of town. More tanks moved in, AT guns, engineers, the AT mine platoon to sweep the crossroads and lay our own, Company 'F' arrived to help hold. Regimental commander on radio put 'E' Company commander in charge so we called it 'Combat Team X' for the moment. CP was the deepest cellar in town, defensive barrages and concentrations all laid out and all tied in for the night. The blasted houses blazed on and Kraut snipers made any movement in town rough. Over 200 PWs sent in by our count. Company 'E' casualties about 90. Started with approximately 170 men, ended with 82."

The Krauts saw the GROSSHAU action from the other side of the fence. The diary of a German soldier who was taken prisoner describes it as follows (The author of the diary, now a PW, had one term of medical school, was a first aid man in an infantry company, in the 1st Battalion, 1058 Inf. Regiment. The PW showed good bearing and is of better than average intelligence):

"**26 November.** 44—Sector GEY—GROSSHAU. (From captured diary)
The hours pass slowly and as I peer out of my hole, the first

dim light shows in the East. The hour is approaching. We expect the Ami (German colloquial for American) to attack at 7:30. Then our fate will be decided, Our CO is looking towards GROSSHAU, he has a good vantage point.

"It's Sunday! My God, today is Sunday, 27 Nov.

"With the dawn the edge of our forest received a barrage. The earth trembles, the concussion takes our breath. Two wounded are brought to my hole, one with a torn up arm, the other with both hands shot off. I am considering whether or not to cut off the rest of the arm, I'll leave it on. How brave those two are. I hope to God that all this is not in vain.

"To our left machine guns begin to clatter—and here comes the Ami. In broad waves you can see him come across the field. Tanks all around him are firing wildly. Now the American artillery ceases, and the tank guns are firing like mad. Can't stick my head out of the hole—finally here are three German assault guns. With few shots we can see several tanks burning once again. Long smoke columns are rising towards heaven. The infantry takes cover, the attack slows down—it's stopped!

"Unbelievable, with this hand full of men we hold out against such attacks. And now we go forward to the counterattack. The Captain is leading it himself. Can't go far, though, our people are dropping like tired flies. We have got to go back and leave a whole number of dead or wounded. Slowly the artillery begins its monotonous song again. Drumming, drumming, drumming, without let-up. If we only had the munitions and heavy weapons that the American has, he would have gone to the devil a long time ago. But as is, there is only a silent holding out to the last man. Our people are over-tired. When the Ami really attacks again, then he has got to break through. I can't believe that this land can be held any longer. Many of our boys just ran away, can't find them, and have to hold out with this small group. But we are going to fight!

"The artillery is still coming in. In the afternoon, all of a sudden, the Jabos (fighter bombers) appear. They are circling over our sector and are looking for targets. Then all hell breaks loose. Not on us, but on GROSSHAU about 200 yards to the left. The motors scream, their MGs rattling, bombs are whistling and detonate with a hellish din. An infernal concert. Then quiet, "I may have to go across to the Amis to ask them for permission to take back wounded and dead. The CO decides against it. Under continued artillery barrage we try to get some sleep. What will happen tomorrow? The Ami is sure to attack again, but can we hold them this time?

"**27 November.** 44—Sector GEY—GROSSHAU.

"I slept like a dead man. At 8:30 it starts again. As usual; first the wild artillery barrage, then the quiet before the attack—an unholy silence before the storm. And then amidst yells they are breaking out of the forest again amongst their tanks. Now our weapons begin to speak. The AMI is getting closer now, but the murderous fire of our MGs forces him to the ground. More tanks are knocked out, and when the AMI sees this, he turns around and retreats into the forest. But the Americans are insulted. Now he really starts hammering us with his heavy weapons—with every blast it seems like our holes will cave in. But if we get hit, it's just tough luck. The wounded are coming in again.

"In the course of the afternoon, the artillery barrage becomes an artillery storm and lasts until night.

"**28 November.** 44—Sector GEY—GROSSHAU. (Last entry)

"The night was pretty bad. We hardly got any sleep. And

in the morning the artillery is worse than ever. Can hardly stand it. And the planes are here again. And once more the quiet before the storm. Suddenly the tanks; and then hordes of AMIS are breaking out of the forest. Murderous fire meets them. But he does not even take cover any more. We shoot until the barrels sizzle. And finally he is stopped again. We are glad, and think that the worst has passed, when suddenly he breaks through on our left. Hand grenades are bursting, but we cannot hold 'em any longer. There are only very few of us left. And there he is again. There are only five of us. We have got to go back; already we can see the brown figures through the trees. As they get to within 70 paces, I turn around and walk away. Very calmly, with my hands in my pocket. They are not even shooting at me, perhaps on account of the red-cross on my back.

"On the road to GROSSHAU we take up a new position. We can hear the tanks come closer. But the AMI won't follow through his gains anyway. He is too cowardly for that.

"Maybe I'll get out of this alive. If so, I can tell the story myself. But if I remain in these torn-up woods, then perhaps a comrade will find this book and send it to my wife."

(This PW was captured by the 'cowardly' American Infantry *who did follow up his gains.*)

Extracts from the 22nd Inf. Unit Journal emphasize the importance, urgency and tensions involved in the Grosshau action during the 28th, 29th and 30th of November.

DATE: November 28, 1944
WEATHER: Cold and Cloudy
PLACE: 025383

Time | To | From

1205 | Red 6/3 | S6
We are still being shelled, one of their aid men turned over our wounded to us, and the aid man (German) came back with them. Believes the Krauts in town have orders to stay until the last man. Have the tanks, TDs and ATs improved their positions? (S6 asks). Not very much. Every time they move, mortars open up. 'B' Company Commander to be highly commended and written up. Asks Red 6 strength of 'B' Company. Thinks its 38 EM and 3 officers, 'E' Company the same. S6 is sure that we'll go in tomorrow. Wants 'B' Company pulled out tonight, 'E' Company to take over. S6 may get some more strength to 'E' Company.

1215 | S3 | S6
Tells S3 of his arrangements with Red 6. Wants White 6 notified of these details. S5 tells S6 that White's large number of battle fatigue consists of many men rested for 24 hours. All in all battle fatigue cases are very low. KIA are tragically high.

1258 | White 6 | S5
We can make officers on the spot. If you have outstanding non-coms will check and put them through if possible.

1330 | G3 | S3
Gives G3 our plans for tomorrow. Tells G3 that we must Take ridge in order to maintain any security. We are Informed by PWs they are in reinforced cellars and artillery

1450 | S3 | Blue
Killed 14 Krauts, got 9 more prisoners. Took over dug-in cannon, going to turn it around. (Task force accomplished this.) Met the lightest

opposition yet in this sector. Our flank move has them fuddled. Fighter bombers claim 3 guns, 2 burnt.

1550 | S3 | Blue 5
Can we hold our turkeys and feed the men a real dinner? No, you'd better make up sandwiches and feed your men.

1600 | S6 | S3
Several engineers were wounded. AT gun crews came to the rescue, killed several Krauts (4) and saved the day. AT lost one man.

1605 | S3 | White 6
Col. Kenan doesn't like the distance between himself and Blue. Trench foot is not doing White Bn. a bit of good; it is generally approaching serious proportions. Got blankets up tonight but always have to leave them. A light tank with trailer will alleviate this situation.

1910 | S6 | Red 5
Red 5 called S6 to give identification of man who committed heroic act. Red said man was Macaro Garcia, 38246362, a PFC and acting Staff Sgt. of 2nd platoon of 'B' Company at the time. S6 said to get two witnesses to act because he is going to put him in for the Congressional Medal of Honor for his heroic acts. (Note: It was awarded after the end of the war).

2025 | S6 | G6
G6 asked situation. S6 said good. S6 told G6 of act of PFC Garcia. G6 said we have a fighting AT Company, and it is a sample of American fighting spirit. As for tomorrow G6 said our attack will be between 4 plus 2 and 4 plus 4.

2218 | White | S3
S3 said that Red 6 is going to handle 'E' Company tonight. Red 6 will turn 'E' Company over to White after first light tomorrow. If there should be a counter attack first thing in morning, 'E' Company will be handled by Red 6.

DATE: November 29, 1944
WEATHER: Rain
PLACE: 025383

0912 | S6 | G3
General Barton wants you to attack three hours, 45 minutes later than you had set a couple days ago. General Barton then came on phone and said the whole front would go today and the 22nd wouldn't be alone.

0912 | S6 | G3
General Barton wants you to attack three hours, 45 minutes later than you had set a couple days ago. General Barton then came on phone and said the whole front would go today and the 22nd wouldn't be alone.

1005 | Blue 6 | S3
The earlier you get started the better, Red will protect your flanks and keep your supply road open. 'E' Company will probe on into town to see what's there.

1010 | S6 | G3
Report that there is direct fire from Grosshau.

1020 | Cactus 2 | S2
According to PW 2 battalions of paratroops were moving into Grosshau last night.

1025 | S2 | IPW
According to PWs they report that if they lost the high ground east of Grosshau or the main road from Grosshau to Gey it will open the way for us to the Cologne plains. Col. Chance is talking with Col. Lanham.

1050 | Blue 5 | S3
The leading elements of the CCR is in the NE of Kleinhau. The Infantry is in the town and mopping it up. Last we heard they lost only one tank. Blue 5 said he had no communication at the moment with Blue 6 but Blue 6 had expected to jump off around 1030.

1106 | White 6 | S3
The Colonel doesn't think you will move today. When you do, you will probably move where Blue is now. The town will be cut and surrounded and when it is, you are to take a good speaking German and go into town tomorrow with a white flag. That is if they continue to hold out. Tell the Commander that HURTGEN and KLEINHAU have fallen - they are surrounded and the hills east and NE of Grosshau are in our hands. Also that 150 surrendered in Hurtgen.

1115 | G3 | S6
Blue got off according to 44th FA at 1100. They had planned on going at 1030 but enemy arty held them up. G3 said that General Barton wants the town (Grosshau) taken today.

1118 | S3 | S6
Orders are that town must be taken at once. Have 'E' Company go in under 2nd Battalion's command. Get the TDs and other support on it.

1120 | White 6 | S3
Order is to take town — use 'E' Company to go in — tanks and TDs

in support. Red will also help to support you. Two (2) tanks and two (2) TDs are to go with them and support 'E' Company under your command.

1158 | S3 | White 6

'E' Company is going to start into town at 1245.

1210 | S3 | Cancan

We are going to fire preparation for White at 1230. 300 rounds into town.

1215 | White 6 | S6

You may be able to use White flag plan. Also use the rest of tanks and TDs if you can. White 6 said one of tanks didn't get back.

1220 | 709 Tank | S6

Find out what happened to that one tank. Also get Co. those other 4 tanks to White. This attack must go on as it has been ordered by the Corp. Commander.

1310 | White 3 | S3

'E' Company jumped off at 1250 at 055389—no report on progress yet. Blue was located.

1327 | S3 | S6

'E' Company receiving counter-attack on its flank. Using tanks in main body to repulse it. Keep Red informed.

1340 | S6 | S3

Counterattack being handled by 'F' Company.

1402 | S3 | White

Leading elements of 'E' Company at 048382 getting heavy MG fire. Think we will be able to take care of situation.

1450 | Red 6 | S6

Get some MGs on north flank of town. 'E' Company is held up about 200 yards inside of town. You will have to move up there fast as reports we get from Blue say they are too far north and will be out on a limb.

1456 | S6 | White 6

We have some 709 Tank people up here. Can I use them? S6 said yes, go ahead and use. Lost (2) tanks on south side by direct fire weapons. These were hit on south side of town. Having quite a fight with 'E' Company. The tanks finally got started and lost two tanks. Here is a warning order: You will have to make plans to take the town from the north tonight if things don't go better with 'E' Company. I don't want you to take too many PWs. If they want to fight to the death, then see that it is their death.

1635 | White 6 | S3

White 6 said that 'E' Company is 75 yards past first house. Plenty of MG and direct fire weapons. S6 said to use 'F' Company to help 'E' Company clear Grosshau. Work until it's taken if you have to go all night.

1745 | S3 | Tanks

Have tanks in town.

1810 | S3 | 46 Ar. Bn.

The 46th Armored is attached to us as of 1630

1815 | Red 6 | S6

Orders Red 6 to aid Blue 6 in every way, shape and form. Watch out for counter-attacks, particularly at dusk and early tomorrow morning.

1815 | S3 | S6

Engineers must clear road through Grosshau - priority job - The 46th Armored has passed the engineers. Get Engineers on that road and make the road passable.

1843 | White 6 | S6

What's the dope on 'E' Company? They were to have three more houses to go and getting lots of prisoners.

1850 | SI

The prisoners are a conglomeration of all kinds of units. They are from 1056-57-58 and parachutists.

1917 | S6 | White 6

The entire town is ours. We have TD on the eastern end. The road will support traffic. 'E' Company suffered heavy casualties. 'F' Company is fair shape. 'G' Company also in good shape. Will have to attack tomorrow at about 1100. S6 congratulates White — Col. Kenan.

1920 | Cactus 6 | S6

S6 informs G6 that the Town of Grosshau is ours. The General says we are doing the job we came here to. "We are whipping those Krauts — we are paying for it — but we are whipping them."

2035 | Blue | S6

Will have more dope for you on the 46th and White later tonight. The casualties are terrific. They have me depressed. Can't believe that over

100 Krauts were still in the town after being pounded with 240 M. M. White phosphorous, etc. It seems incredible.

2100 | S6 | S2
We have a total of 107 prisoners so far. We have reports of paratroopers. We can expect them in the woods from here on.

2120 | S6 | G6
YOU broke the Krauts today and while you'll have to move tomorrow, I'm (G6) sure it won't be difficult.

2125 | S3 | Blue 5
Want that road swept faster. It will be a slow job says S3 as there are plastic and glass mines to be sweated out. Your best bet is through the center of town where tanks and TDs ran.

2125 | White 6 | S3
You'll have the 46th on your right. Keep on your left, with all the tanks and TDs tomorrow. The attack will start at 1100.

2245 | S3 | White
At 058383 troops observed digging in. Note long overcoats and long shovels observed by 'E' Company. S2 believes they're friendly troops.

During the night of **December 1-2,** all available personnel of Regimental Headquarters, Antitank, and Service Companies were organized and alerted as reserve riflemen. The Combat Team's fighting strength now consisted mostly of replacements who had joined during the past fourteen days.

THE 22ND INFANTRY JOURNAL

DATE: 1 December, 1944
WEATHER: Fair
PLACE: 025383

Time | To |From

2130 | S6 | G6
G6 said there will be a reinforced company on hill with TDs and ATs and possibly tanks. G6 said to go ahead with S6 plan. S6 said 'A' Company's platoons are commanded by Pvts. of three days service. G6 said he understands the situation. S6 said in order to do this thing tomorrow he has scraped the bottom of the barrel. S6 said he needs Cargo to tie in to us tomorrow. G6 said Cabbage made about 700 yards today. G6 said that General Collins told him that the 22nd Inf. has been fighting a wonderful fight under severe handicaps. G6 said the outstanding performance since D-day has been by the 22nd Infantry.

22nd Infantry in the Hürtgen Forest, Germany December
1, 1944 — Photo from the National Archives

Jeep of 2nd battalion 22nd Infantry in Grosshau, Germany
December 1, 1944 — U.S. Army Signal Corps photo

*9 — 22nd Infantry mortar team in Grosshau,
Germany — U.S. Army Signal Corps photo*

HEADQUARTERS 25TH INFANTRY REGIMENT
FORT BENNING, GEORGIA

E-X-T-R-A-C-T   7 May 1946

GENERAL ORDER — War Department
NUMBER 37
Washington 25, D. C.

19 April 1946

IX — BATTLE HONORS. As authorized by Executive Order 9396 (sec. I, WD Bui. 22, 1943), superseding Executive Order 9075 (sec. Ill, WD Bui. II, 1942), the following units are cited by the War Department under the provisions of Section IV, WD Circular 333, 1943, in the name of the President of the United States as public evidence of deserved honor and distinction. The citations read as follows:

1. The 22nd Infantry Regiment, with the following attached units
   Company C, 70th Tank Battalion
   Company C, 803d Tank Destroyer Battalion
   Company C, 4th Engineer Battalion
   Company D, 70th Tank Battalion

is cited for extraordinary heroism and outstanding performance of duty in its determined drive to overcome bitter opposition in the Hurtgen Forest. On 16 November, the 22nd Infantry Regiment, with attachments, opened an offensive with the mission of clearing a portion of the Hurtgen Forest of powerful enemy forces and fighting its way to the Roer River and Cologne Plain. Throughout the campaign, the progress of the unit was seriously impeded by an unusual combination of inclement weather and difficult terrain. Unseasonable precipitation and damp, penetrating cold were a constant detriment to the health and well-being of the personnel. The terrain was characterized by densely forested hills, swollen streams, and deep, adhesive mud, which retarded all movement of troops and vehicles. Fully cognizant of the decided strategic advantages which this area afforded for effective defense, the enemy had pre-pared an elaborate system of mutually supporting fortifications. The effectiveness of enemy artillery and mortar fire was considerably enhanced by the frequency of tree bursts in this heavily timbered area. Inasmuch as natural conditions and rigid construction of enemy strong-holds frequently precluded the effective employment of aerial and motorized support, the burden of neutralizing fanatically defended enemy fortifications fell heavily upon the shoulders of the infantrymen, as exemplified in the capture of Grosshau, a town in which concrete and steel reinforced basements rendered each house veritably impregnable to repeated artillery and ae-rial attacks. The town was ultimately captured by an assault in which the infantry closed with the enemy in hand-to-hand night fighting. The 22nd Infantry Regiment with attachments, cleared its portion of the Hurtgen Forest and reached its objective on 4 December 1944, opening a gateway to the Cologne Plain and the ultimate rapid conclusion of the European conflict. The in-dividual courage, valor, and tenacity displayed by the personnel of the 22nd Infantry Regiment, with attachments, in the face of superior odds, unusually hazardous conditions, and unfavorable weather were in keeping with highest traditions of the armed forces.

By Order of The Secretary of War:
OFFICIAL: DWIGHT D. EISENHOWER
EDWARD F, WITSELL Chief of Staff Maj General The Ad-jutant General
ROBERT P. STRICTLAND A TRUE COPY Capt. 25th Infantry Adjutant

* * * * *

Life Magazine reported the last few days action of Hurtgen For-est in its Jan. 1, 1945 issue as follows:

THE BATTLE OF HURTGEN FOREST

A GLOOMY GERMAN WOODS TAKES ITS PLACE IN U.S. HISTORY BESIDE THE WILDERNESS AND THE ARGONNE

By William Walton

"Five miles southeast of Aachen is the Hurtgen Forest, 50 square miles of tall firs and Siegfried line pillboxes. In September, U. S. troops went into the forest and after ten weeks of eerie, murderous fighting they came out of it. Last week the German counter-offensive threatened to outflank the American holding the forest but told by "Time" and LIFE William Walton, it could not efface the story of courageous men who took it.

"The sergeant named Garcia couldn't believe his eyes when he saw six American engineers warily working their way across the flat, unfenced fields into Grosshau. Garcia knew that Grosshau was still very much in German hands. As he watched in the cold gray afternoon from his shell crater he saw happen just what he knew would happen.

"The Germans burrowed into the cellars of the ruined village, let the engineers creep 200 yards past the first house, then opened fire from all sides. The engineers disappeared in a burst of flame, either dead or prisoners.

"Nobody could find out who had given the engineers orders to enter Grosshau. One of those snafus which are part of any battle. Neither could anyone find out, after the battle of Hurtgen Forest had ended, just exactly what Sergeant Garcia had done then. All they knew was that Garcia and a half dozen other gunners had left skeleton crews at their antitank guns a quarter of a mile up the road from Grosshau. Then they crawled down into the village outskirts, killed the Germans in the nearest cellars, recaptured the engineers, and crawled back with them to the American lines.

"Garcia was wounded. So nobody could get a detailed account from him before he was carried back to a clearing station. Not even his first name. Somebody remembered he had just got American citizenship a year ago. That was all. The regimental colonel when he heard about it said, "That man is going to get the best medal I can give him. Somebody must find out all the details." But the fight for Grosshau and the last dank patches of Hurtgen Forest were still too hot just then for any careful research, and Garcia's feat had been only a little more heroic than those of hundreds of other men who in the tumult and confusion had been daring and courageous and resourceful. Most of their braveries never would be known, except to the few who witnessed them, just as no man would ever know all that had happened in the battle of the Hurtgen Forest.

"Hurtgen Forest is a name to carve some day on the war memorials of America beneath such evocative place names as Chateau-Thierry, the Argonne forest and the Wilderness of the Civil War. Other battles in this war have been more dramatically decisive—Normandy, St. Lo, the Falaise pocket—but none was tougher or bloodier than the battle for this Hurtgen Forest.

"Close-ranked firs towering 75 to 100 feet make the Hurtgen Forest a gloomy, mysterious world where the brightness of noon is muted to an eerie twilight filtering through dark trees onto spongy brown needles and rotting logs. Occasionally a neat ditch to control forest fires slices through the overgrowth. A few woodchoppers' huts such as old Germanic folk tales describe are hidden among the trees. On the western fringe four villages—Rott, Zweifall, Yicht and Schevenhutte—fill small clearings. Otherwise the Hurtgen-wald is a fathomless sea of darkness, somber enough in peacetime, in wartime sinister with lurking enemies, evil with whining bullets and bursting shells that leave broken trees and broken men in tangled fraternity.

"American troops had been in the forest since mid-September when the 1st, 4th, and 9th Infantry Divisions overran the outer Siegfried defenses with the momentum gained through France and Belgium. Two Hurtgen villages, Zweifall and Schevenhutte, fell before over-taut supply lines pulled the First U. S. Army up short. Not until mid-November could the starting signal be given again.

"By Nov. 16, the 9th Infantry Division had been withdrawn from its dug-in Hurtgenwald positions and the 4th Infantry Division, first ashore in Normandy, substituted to spearhead the attack. The 4th's commander, Major General R. O. Barton, sent his 12th Regiment in on the left when the offensive started. Two days later, the 22nd Regiment was committed in the center, then the 8th Regiment on the right. The Germans brought to bear artillery and mortars in concentrations such as had not been heard nor felt previously on the Western Front. Pine needles hid vicious Teller mines, box mines and the new shoe mines which blew unwary patrols up into branches minus a foot or a leg or a life.

"The 12th Regiment, under hoarse-voiced Colonel Bob Chance, worked foot by foot up a forested slope with two companies, 'F' and 'G', driving a wedge into the German lines, a wedge that threatened the German positions but also exposed the two companies' flanks. The German mortars, wise to the terrain, cracked into 'F' and 'G' Companies, bursting in the trees to shower jagged fragments for yards around. Machine guns ripped the gloom and rifles crackled as the Germans gave a little ground, but only a little.

"Seeing the two companies worm into their lines, the Germans waited. Then they threw in concentrated mortar fire and under its protection struck down a ravine and up the other side to cut the slim supply line. 'F' and 'G' Companies were trapped.

"For two days and nights the Germans poured mortar and artillery shells into the narrow area where infantrymen back to back were fighting off the German attacks. One slender footpath brought a trickle of ammunition but no food or water. The path was under such constant fire that the wounded could not be evacuated. Rain drizzled through the darkness, trickling into foxholes, and seeping through winter clothing. Medical supplies were in-sufficient to care for men with jagged leg wounds, with bleeding chests, missing fingers, blood-and-rain-soaked bandages.

"'E' Company, nearest to their surrounded comrades, tried desperately to relieve them. On the third day, Colonel Chance moved up 'A' and 'C' Companies in the darkness, sent them in to attack at daylight. The Germans, caught between the isolated companies and their relief, were slaughtered. On both sides there was slaughter, but 'F' and 'G' Companies had been saved. That was how every foot of the Hurtgen Forest was to be.

"Then it snowed. A wet, suffocating blanket sifted through the trees, weighting heavy-branched firs, covering foxholes and shell craters, mantling sentries who stomped at their posts and patrols creeping under low boughs. The few roads, already pale brown rivers, remained quagmires that sucked down tanks, trucks, and jeeps struggling toward the front.

THE 22ND SEES LIGHT THROUGH THE TREES

"Now the focus of battle shifted to the 22nd Regiment whose commander, Colonel Buck Lanham, could see light ahead through the trees. His battalions were spread like two fans on either side of the ravine. Muddy trails from the rear brought up replacements to fill out his riddled lines, fresh-faced second lieutenants still untried by enemy fire, privates in clean overcoats standing in the trucks with grave expressions, not scared but very serious as the sound of shelling echoed through the trees. In his wooden trailer, Colonel Lanham was talking on a field phone: "Tell 'C' Company to get in there with bazookas and grenades and take that high ground. Hill 90. We've got to have that high ground. Hill 92 doesn't do us any good until we've gotten 90, too. And Blue can't advance until the Krauts are knocked off those hills. Tell them to fight like hell." He hung up and turned tensely to the map on his small folding table. Prematurely gray, with black eyes bright behind his spectacles, the colonel seemed too absorbed in the map's crayon hieroglyphics to notice the stocky, wide-faced captain waiting across the table. Without looking up he said, "You didn't know it was me you had on the phone last night, did you, Swede?"

"No, sir," said the captain, shifting nervously. When the colonel looked up he smiled. Swede smiled too, with relief.

"That's all right, Swede," the colonel said in a voice softened by understanding. "I know how it is when you see a lot of your friends knocked off. But you've got to treat your superior officers with more respect."

Swede was silent a moment, then he said quietly, "Colonel, sir, I don't care if you break me for it. I meant what I said last night, even though I didn't know it was you on the line. That little patch of woods we're fighting for ain't any good to anybody. No good to the Germans. No good to us. It's the bloodiest damn ground in all Europe, and you make us keep fighting for it. That ain't right."

Now it was the colonel's time to be silent. Two men sitting across the table looking at one another in silence. The colonel, slight of build, keen-faced, intense. The blond captain, bulky, mud-spattered, a two-day growth of beard on his wide face, a face designed for grinning but dead serious now and pale with fatigue.

THE BATTLE IS FOR KILLING GERMANS

"There's nothing in the world," said the colonel deliberately, "that I'd like to do better than tell all you boys to call it off and go home. You know that, Swede. But it can't be done. The only way we can get this thing over is by killing Krauts. To kill them you've got to get to them."

Swede grunted.

"Look here on the map. You know they're dug in all through this woods you're talking about. Once we've got those two hills, then we'll be able to pour so much stuff into that patch of woods that not a Kraut will be left. Then we can push on to where the woods end and fight in daylight like little gentlemen again. Wherever there are Krauts we've got to kill 'em. I know they've killed lots of our boys in that patch, but we've killed even more of them and that's what counts."

Swede sighed. "I know you're right, Colonel. Knew it all the time. I just have to get things off my chest once in a while."

"Pour yourself a slug of good Heinie cognac," said the colonel. When Swede left, he was smiling again. The phone buzzed.

"Charcoal speaking," said the colonel. "Yes, wild man. Got up on top the hill, did they? That's the stuff. Keep 'em going and let me know so we can start concentrating on that wooded patch."

No sooner had he hung up on his G-2 officer than the phone buzzed again.

"Charcoal speaking. Oh, hello, General. Yes, Blue is going to jump off soon. Just as soon as Hill 90 is cleared. Yes, sir. That's right. You can count on us, but I wish you'd get that damn task-force armor to start moving south of us. You know how armor is, wants every foxhole cleared out before they'll move.... Well I suppose so.... Yes, sir."

During the dreary afternoon, Lanham's 'C' Company fought up to the top of Hill 90 and started firing down into German positions before the early winter twilight made them button up for the night.

A colorless dawn brought more artillery, but by 8 o'clock the gray sky had broken sufficiently for fighter-bombers to lay on a mission. Tobogganing out of the western sky they came down over Grosshau, the tiny village just beyond the eastern edge of the forest. Each cracking explosion fountained smoke and debris into the still air. In foxholes scooped from rotting pine needles the foot soldiers watched approvingly. "Give it to the bastards." "Now we're getting somewhere." "Lookit those houses go whammo."

In the thunder heaped on Grosshau it seemed impossible for any living thing to survive. Cautiously riflemen and Tommy gunners hunched down the hill from tree to tree, firing whenever a shadow moved unnaturally in the woods ahead. Now they were fighting in Swede's bloody patch of woods, where every tree was shattered into a naked spear of white ugliness against the dark earth, where weather-soaked corpses had lain so long the stench was unbearable.

As the day advanced, the dirty brown uniforms drove the dirty gray uniforms out of the last woodland west of Grosshau. Emerging onto treeless ground, the Americans felt as naked as stripteasers at a Sunday-school picnic. The forest which had been hateful seemed friendly and protective now that they had only tiny hillocks and shell craters to shield them.

Artillery hidden in the woods across the barren plain intensified their thunder when the Americans appeared. From crater to crater Lanham's men dodged, fired over rims, ducked to another crater farther ahead. From Grosshau, which had seemed completely uninhabitable, the machine gun and small arms fire grew intense.

Time after time the 22nd pushed forward, stumbled, took cover. Six Sherman tanks were knocked out. The lines regrouped to try again. But too many men were falling. The plague of shells grew even thicker.

Back in the forest, a major talked to Colonel Lanham by phone. When the major had explained, Lanham said, "All right. Have your boys dig into the best positions possible. Just hold on. We'll have to try it another way."

The other way was to keep to the woods and circle northward around Grosshau and the plain. Already Major George Goforth's battalion was pushing over a series of wooded hills that bulged like a ripple of muscle along the forest rim. That was on Monday, the 27th. For two days they fought through those hills, with casualties bad on both sides.

In his gloomy CP, a lantern-lit log dugout, Major Goforth talked over the situation with his exec, Swede Henley. A company commander had just come in to report.

"We're hunting for officers, " said the new arrival, slumping onto a battered tin water can. 'G' Company's got only two officers left. Lost three this afternoon. We can't go on like this, Major," Goforth shook his head.

"I know, boy, but where am I going to get them? Division says we can commission any good man right here in the field. But who?" He looked around a circle of dirty, unshaven faces watching him in the sputtering light, faces drained of color like those of drowned men.

"There's McDermott," said Swede. "Can't spare him. Practically runs G-2."

"He's the last available sergeant. We've already commissioned six." "Guess we'll have to depend on replacements," said Goforth. "The trouble with replacements is that they don't last long enough," observed the company commander. "Trucks brought up 30 for me this morning; 18 were hit even before they could get into the line. No percentage in that."

As he spoke, the blanket covering the dugout doorway was pushed aside and three young lieutenants entered, saluted, and said they were reporting for duty.

"There you are, Jack," said Goforth. "Replacements for you. Take 'em with you when you go back."

THE AMERICANS COME OUT OF THE WOODS

Somehow all that day and the next, Goforth's battalion, despite its losses, managed to hump over the wooded hills, edging closer to the road connecting Grosshau with Gey. Another battalion just held onto their positions a little way outside the forest facing Grosshau, taking losses from artillery, too, but hanging on. The second day was when Sergeant Garcia charged into Grosshau and rescued the six engineers.

During the night, an order came down from higher head-quarters that Grosshau must be taken next day, the 29th. Maybe something in the big picture made it necessary. Nobody knew. But those were the orders, meaning frontal assault on Grosshau. Wearily, the preparations were made.

At 0900 under a low gray sky the first infantrymen raised themselves from their shell holes into machine-gun fire that spurted from the ruined village. Behind them came two M-10 tank destroyers, mounting three-inch guns. The tank destroyers rolled ahead where infantry could not make it through gusts of bullets, rolled near enough to German positions to silence their guns, then on to the outskirts of the village.

Ducking low, riflemen advanced 200 yards behind them, us-ing what cover there was, firing toward the village ruins. When the tank destroyers had rumbled beyond four houses and a jagged fragment of a church, the first break came. Fifty Germans, bleary and dust-covered, scrambled from cellars shouting "Kamerad" above the noise of battle. Platoon Sgt. Stanley Ward from an M-10 turret waved them back toward the infantry.

Then it was a slow, house-to-house fight, warily spraying every doorway and shed with gunfire, hurling grenades into each cellar opening, herding prisoners down the one muddy street strewn with dead men and horses, timbers, bricks, and dirty straw. Nei-ther side was shelling Grosshau, but shells were scraping over-head toward rear road junctions and supply depots.

In three hours the worst of the fight was over. A few snipers lurked in among the ruins, but the muddy Americans were able to push beyond the village, set up mortars and to start attacking German entrenchments in the open field. Grosshau was ours. Between the edge of the woods and the village, 250 Americans had died.

While the fight for Grosshau was at its bitter height, Major Goforth's battalion had crossed the Gey road and pushed over more hills. To the south, Kleinhau and Hurtgen, two other villages, had fallen. Twice the Germans threw violent counterattacks against Goforth's men, but using his reserves and cooks, guards, and engineers from Colonel Lanham's headquarters, the counterattacks were stayed. The battle of the Hurtgen Forest had drawn to its end."

The newspapers back home reported the battle as follows:

Chicago Daily News — Nov. 25, 1944.
"Dozen Divisions Hurled at YANKS."

"Paris (UP)—Ten to twelve Nazi Divisions fought four Allied armies to a standstill in bloody showdown battles on the Cologne and Saar plains today." "The great allied offensive on a 25 mile front inside Germany southeast, east and north of Aachen had ground to nearly a complete stop on its 10th day in the face of German counterattacks and last ditch resistance on the west bank of the Roer River before Cologne.

Supreme Headquarters reported "Slow but steady progress" against very strong resistance in the Hurtgen Forest southeast of Aachen, but elsewhere the American 1st, British 2nd, and American 9th Armies appeared stalemated temporarily in toe-to-toe slug fests with German infantry and armor."

Chicago Sunday Tribune — Nov. 26, 1944.

3rd ARMY HITS INTO GERMANY AT NEW POINT

1st at Outer Edge of Hurtgen Forest SUPREME HDQ. AL-
LIED EXPEDITIONARY FORCE

Paris Nov. 25 (AP) The American 1st Army pushing steadily
toward the Cologne plain against undiminished German opposition reached the outer edge of the bloody Hurtgen Forest tonight and was under mounting robot bomb fire from the Nazi
defenders at the Rhine.

The troops fought within a few hundred yards of Grosshau
and a thousand yards of Kleinhau.

Chicago Daily Tribune — Nov. 27, 1944.
SUPREME HDG. ALLIED EXPEDITIONARY FORCE.

Paris Nov. 26 (AP)

Four American Armies smashed forward on the western
front in battles of undiminished fury today.

On the bloody battlefields of the fronts northern sector,
doughboys of the American 1st Army took Weisweiler. An important height taken near Duren is just north of Grosshau, five
miles southwest of the Roer bastion and to reach it the doughboys had to fight across terrain where the Germans had fortified
every house and connected them with trenches which were improved versions of those in the First World War."

Chicago Daily Tribune — Nov. 30, 1944.
"MEETS MONTY ON FINAL BLOW

9th ARMY PIERCES ROER DEFENSES;
3rd nears Saarbruckn. Bulletin

At a rendezvous in Belgium (AP):

Gen. Eisenhower and Field Marshal Montgomery met to-night and planned measures for the final defeat of Germany.

Surging forward in predawn darkness 1st Army Infantry of Lt. Gen. Hodges fought into the streets of Inden and Jungers-dorf and close to Lamersdorf.

Other men of the 1st Army fought in the streets of Grosshau and for the last third of the forest battlefield town of Hurtgen."

"HURTGEN FOREST WRITES ITS
NAME BESIDE ARGONNE"

By William Strand
(Chicago Tribune Press Service)

With U. S. 1st Army troops in Hurtgen Forest, Germany—**Nov. 29. 1944**

Deep within the spruces and pines of Hurtgen forest, south-west of Duren, American doughboys are writing one of the grimmest chapters in the history of this war. When the full story is told of its cost in men and materials, the name of these woodlands probably will rank alongside the Argonne Forest in the First World War and the Battle of the Wilderness in the Civil War.

The 4th Infantry Division, commanded by Maj. Gen Ray-mond O. Barton has fought partway through the cold and rain-soaked woods, pushing the Nazi back yard by yard. Before that

the 28th Division battled its way forward through the thick un-
derbrush against enemy snipers, machine gunners, barbed wire,
and countless mine fields.

REACHING EDGE OF FOREST

Tonight, dirty, tired infantrymen were nearing the eastern edges
of the forest at several points.

You can't measure this fighting in yards, or towns taken; any
more than you can measure it in the number of cases of swollen,
trench feet that have developed, or the length of the beards the
men have grown because they were too busy staying alive or too
tired to shave."

Chicago Daily News — Nov. 30, 1944
"STORM ROER LINE London (AP)

The Roer River line was cracking today under the assault of
275,000 Yanks in 23 divisions.

The 1st Army at the southern end of the critical 25 mile front
on the Cologne plain captured Lamersdorf and Grosshau and
emerged from the Hurtgen Forest pinelands."

* * * * *

The next day, the 30th, brought forth a new task. An attack was
to be launched to the northeast by the 2nd and 3rd Battalions.
The 46th Armored Battalion, attached to CT 22 the day before,
was to be abreast of the 2nd Battalion and on the extreme right.
The objective was the edge of the forest 1,000 yards to the north-

east of Grosshau from which a coordinated infantry-armor attack could be made on Duren.

After a heavy artillery preparation, the 2nd and 3rd Battalions attacked. At first, the 3rd Battalion moved against comparatively light resistance and by 1500 hours was only 300 yards short of the objective. The opposition became intense, and the Battalion was forced to secure for the night.

The 2nd Battalion had moved approximately the same distance across the open plain against enemy entrenched and called for armored aid. Four tanks and two tank destroyers were sent up, but intense shelling and small arms fire held up the unit on the open fields.

The 46th Armored Battalion, in an approach march to the line of departure, received small arms fire from the high ground northeast of Kleinhau. This hill had been reported secured the previous day by elements of Combat Command 'R' of the 5th Armored Division. The Battalion was forced unexpectedly to fight across open ground and assault a fortified hill. By 1615 hours the battalion was at the woods southeast of Grosshau and had only an estimated 50% of its effective fighting strength left.

The 1st Battalion, in reserve, moved up to positions in rear of the 3rd Bat-talion. During the advance to these woods, Companies 'A' and 'C1 had protected the flanks and rear and had given fire support. Company 'A' was following up behind Company 'C' and as Lt. DON WARNER, 'A' Company Commander, later signified, "'C' Company had apparently barreled through the wooded area without too much trouble, but they stirred up the commotion which greeted us."

Positions on the 1st of December were as follows: The 3rd Battalion and the 2nd Battalion on line, the 3rd on the left some

1, 200 yards northeast of Grosshau, and the 2nd about 700 yards southeast of that town. The 1st Bat-talion, in reserve, controlled the hill 400 yards northeast of Grosshau and covered the 800 yard gap between the forward battalions.

The missions were: the 3rd Battalion would attack northeast, securing the woods southwest of Gey; the 1st Battalion from the rear of the 3rd Battalion would attack southeast, hitting the flank of the enemy facing the 2nd Battalion; and the 2nd would stage a holding attack. Having disrupted the enemy's de-fenses to the front of the 2nd Battalion, the 1st then would secure the woods edge abreast of the 3rd Battalion, and the 2nd Battalion was to advance to the northeast.

As planned, the 3rd Battalion attacked at 0900 hours and within two hours had pushed to its objective and had organized a defensive position. Screened by smoke, the 1st Battalion's attack to the southeast at 1000 hours was highly successful. It surprised the Germans, and rapidly the attack moved to the woods 500 yards northeast of the 2nd Battalion. Leading elements of the 1st reached the edge of the woods west of Strass by noon and tied in with the 3rd Battalion to the north. The 2nd Battalion with com-panies 'F' and 'G' continuing the attack to gain the woods east of Grosshau encountered very-stiff resistance but by late afternoon had accomplished its mission.

The diary of Capt. Faulkner, Company 'E', then in 2nd Bat-talion Reserve, comments on the **December 1-2 actions.**

"Dec. 1, 1944. Made recon with runner in the open during the morning for possible attack. Jumped in hole to miss arty and said good morning to dead Jerry machine gunner left from day before still sitting behind his gun. Company 'E' held Grosshau. Had pictures taken while in our CP. Were told would be in Yank and Stars and Stripes. Also heard Ernest Hemingway who was up at Regimental CP was going to write up Hurtgen Forest. Late afternoon got a rush order to go to assistance of Company 'F' who were in a bad counter-attack. Sent 1st platoon (14 men under Sgt.) over the open field alone. Saw it pinned down under timer fire, mortar fire, and MG fire from right flank hills. Wish we had smoke. (9 men got through.)

"Received more news of 'F' being attacked by Jerry and in bad. Now dust but moved Company at the double long way around through woods and across open. Flushed some Krauts and passed them back down the Company line. Wonder where those PWs ended up?

"Lt. Mason and myself ended up lead scouts just to get there. Found 'F'. Lt. Wilson and 15 men left out of 150. Said Lt. Fitzgerald was the hero of the day. 'F' Company moved back at midnight. We made too much noise cutting trees for our holes and some replacement shouted "Timber". We got a mortar barrage for it and more casualties.

"Held all next day. Dec. 2 under increasing enemy arty and mortar barrages. Casualties increased. Lost my two best Sgt. platoon leaders, Ivy, and Richardson, both wounded. Too much attrition.

"Registered own artillery and mortar fire all day for protective barrages. Tried to find Jerry OP and knock it out with fire. Warned all men of possible Jerry counterattack that night or at dawn and skinned in and out of holes be-tween bursts. Could see Duren through my glasses down in the open valley,

"Dec. 3, a great day. We are to be relieved by 'K' Company 83rd Division. A bad day for enemy arty and mortar fire. Suffered 3 direct hits from light mortars on CP log roof but no casualties in our hole. Radio warning 'Watch for enemy air', Hell! They're over us right now. Strafed us and bombed 46th armored

Inf. 1000 yards to our right rear, (S. W.) Came in 3 sweeps. Terrifying. We are relieved and start pulling out of our position only to be covered by bursts of Kraut MG fire. I pray hard and call for our concentration 45 Baker on ridge to our right. One or the other does the job as Jerries fire dwindles to almost nothing. The rest is easy. It's dusk as we walk back in the driving rain through the deep muddy fields and on through Grosshau. This part is over. Company 'E' had 79 men and 6 officers before Grosshau, received about 160 new men and now only a little over 80 with 3 officers are go-ing back. It had better be worth it. "

The 46th Armored Battalion, still making gains against stiff opposition but severely depleted in strength, was ordered back to secure previously held positions and was detached from the CT at midnight.

During the night of December 1-2, all available personnel of Regimental Headquarters, Antitank, and Service Companies were organized and alerted as reserve riflemen. The Combat Team's fighting strength now consisted mostly of replacements who had joined during the past fourteen days.

THE 22ND INFANTRY JOURNAL

DATE: 1 December 1944
WEATHER: Fair
PLACE: 025383

From

G6

G6 said there will be a reinforced company on hill with TDs and ATs and possibly tanks. G6 said to go ahead with S6 plan. S6 said 'A' Company's platoons are commanded by Pvts. of three days service. G6 said he understands the situation. S6 said in order to do this thing tomorrow he has scraped the bottom of the barrel. S6 said he needs cargo to tie in to us tomorrow. G6 said cabbage made about 700 yards today. G6 said that General Collins told him that the 22nd Inf. has been fighting a wonderful fight under severe handicaps. G6 said the outstanding performance since D-day has been by the 22nd Infantry.

Th poe plan for the **2nd of December** was for an attack to be made to the south by elements of the 1st and 3rd Battalions to clear a tongue of woods east of the 2nd Battalion and assist the advance of that unit.

However, at 0650 hours a strong German counterattack by an estimated battalion of 250 fresh infantrymen broke through near Company 'I' and advanced toward the high ground northeast of Grosshau. During the operation, Lt. McCahill, who had only hours before assumed command of Company "I" when Lt. McCollum, the only remaining officer in the company was seriously wounded in the hip, was captured by a platoon of Germans. When captured, Lt. McCahill, the 44th Field Artillery observer, Lt. Hayden, and about twelve enlisted men were in a re-enforced water tower. They talked the German officer into leaving them in the tower with several guards while the rest of the Germans continued the attack. After a few moments of consideration, the German lieutenant agreed and left them with three guards.

Persuaded by a little fast talking, the guards gradually relaxed and began giving the American prisoners 'K' rations and hot coffee. In the meantime, the Germans were being driven back and Lt. Burton, with a detachment from Company "K", moved forward to seal the German penetration. In so doing they recaptured the water tower and the surrounded group. At 0740 hours, the 3rd Battalion requested that all available men be sent forward. Capt. Tommy Harrison, with as many replacements as could be gathered together, fought his way to the 3rd Battalion OP and helped alleviate the pressure on the 1st and 3rd Battalion Command Posts.

Shortly thereafter, enemy were reported on the hill northeast of Grosshau, and all available armor was ordered to retake the ground. The three 81 mm mortar platoons were firing continuously in an effort to destroy enemy organization and communications. The mortar platoon of Company "M", led by Lt. DAVID JAMES, fired over 1,100 rounds of mortar ammunition during the first thirty five minutes of the counterattack. Artillery and mortar barrages were brought into the very limits of the forward units, often in the gaps between companies.

In one isolated building, Sgt. Wertman directed both artillery and mortar fire directly on top of his OP in an effort to drive the Krauts away. The 1st and 2nd Battalions had shifted their depleted reserves into positions to aid the 3rd Battalion, should a penetration be made. By 0955 hours, the situation had stabilized to a certain extent. The Combat Team continued resistance; replacements were brought forward; the rifle units organized from reserves the night before fought and cleared out infiltrated enemy pockets; and at noon the situation was under control. Front lines were again established on positions held the previous night, with the center line drawn back to take advantage of more favorable terrain. As relief of the Combat Team was close at hand, all units bettered their positions.

* * * * *

The relief of Combat Team 22 by Combat Team 330, 83rd Infantry Division, began at daylight the **3rd of December.** During the relief, an enemy attack hit elements of the 1st Battalion, but this was repulsed, and by 1230 hours the lines were again stabilized.

Shortly after noon and during the height of the relief, approximately forty enemy pursuit planes roared over the front lines. The first impression from the roar of the planes was that it was a ricochet round from a German 88. Before one had time to locate a defiladed area, the planes were on top of the troops. To the other men the grinding motors meant it could be nothing but the enemy. It was a strange coincidence, but the German planes were never seen when our air corps was out; evidently they feared a dogfight and timed their missions accordingly.

As the first plane sped over, it was quite possible to see the pilot, as he skimmed the tree tops. In the distance the little artillery cub planes scooted down out of sight. Men and vehicles quickly dispersed to the first available cover. Several men ran to the jeeps and mounted machine guns to return the fire. Within thirty seconds, one of the planes was set ablaze in the 1st Battalion area and crashed.

The first trip over wasn't costly as the planes were searching for targets, but as they circled back they began their strafing and bomb runs. Antiaircraft guns fired incessantly, and the abrupt puffs of smoke filled the air, and another German plane soared earthward.

As these planes streaked back toward their bases, they dropped the huge anti-personnel bombs along the roads and in the open fields. A personnel bomb was a massive bomb designed to split open when it struck the earth's surface. When it split open, hundreds of smaller fragmentation bombs filled the air with flying shell fragments.

The men huddled in the draws, in shell holes, and in slit trenches. It was extremely dangerous to look about as the planes roared in from every conceivable direction spitting their deadly steel. Bullets ricocheted from all angles, cutting the grass underfoot, snapping limbs overhead, and pock-marking the roads. The aerial attack lasted slightly over thirty minutes, but each minute seemed an eternity.

After it was over, the men wondered how it was possible for the German soldiers to withstand the daily strafing which our attached planes delivered. Later in the afternoon, the silver P-38's, an unmistakable friend, streaked through the atmosphere to avenge the threat to allied air supremacy. These planes were a beautiful sight when the bright sun flickered from the wings and fuselage almost blinding one. As soon as these planes could be detected, the enemy artillery diminished to almost nothing. They were a godsend.

Because of the air raid, mortar fire, and artillery fire, the relief was delayed, and it was not fully affected until dark that day.

Upon relief, the CT moved into assembly areas five miles west of the front lines and staff planning began for the move into Luxembourg.

* * * * *

CT 22 had completed its battle through the Hurtgen Forest. **In the eighteen days of the campaign, the CT suffered 2,575 enlisted men and 103 officer casualties. The 44th Field Artillery in the fighting lost 11 of its 12 original forward observers.** Before the battle was over, battery commanders, battery executives, and the motor officers had served as forward observers. The advance had covered 7,500 yards and pushed the front lines to the open ground which formed the approach to Duren.

And so the battle for Hurtgen Forest had ended; bloodiest, most demanding of all the battles the Fourth had fought to date. It was truly named "The Death Factory." Not at any other time did the troops have to undergo such discomforts. No other battle called for such sacrifices of men and material, nor needed such stamina to carry on.

The fight had been carried on at a white hot pitch from the initial jump-off until the division had been relieved. Never for an instant any time of the day or night had a single member of the Combat Team been free from the petrifying, draining presence of fear and exhaustion. It is incredible to believe that against overwhelming odds and everything which the powerful German Army could throw against them that the Combat Team still fought its way through the forest. PFCs commanded platoons, frequently after only hours of experience in combat.

The Germans themselves could not believe that the Americans could still continue to fight when they well knew the extent of the casualties. And so one of the miracles of battle was finished. Thousands of men had passed through the Regiment in 18 brief days. Hundreds of men had lost their lives for a patch of woods and the heap of rubble that was Grosshau. Perhaps in the final analysis, the sacrifice demanded in Hurtgen will be deemed worthwhile; we wouldn't know about that. We are only the men who fought the battles, who hugged the earth as our hearts dropped within us when the shells came screaming in, who lay in the slime and mud night after night, who froze our feet, who did not come out of a foxhole long enough to eat Thanksgiving dinner, for life was more precious than food, who carried our buddies back to the aid stations or turned our faces aside when we saw what was left of their broken, mangled bodies. A part of us died in the forest, and there is a part of our mind and heart and soul left there. We shall never forget; we are incapable of forgetting. This was War. This was Germany. This was the Twenty-Second Infantry. This was Hell. This was the Battle for Hurtgen Forest and Grosshau.

* * * * *

HEADQUARTERS COMBAT TEAM 22 COMMAND POST OF THE COMBAT TEAM COMMANDER

Grand Duchy of Luxembourg
9 December 1944

To The Officers And Men Of Combat Team 22:

16 November to 3 December 1944 you fought one of the greatest regimental battles in military history. By the end of the sixth day, you had suffered approximately 50% casualties, the point at which a regiment is considered to lose much of its effectiveness as a fighting instrument. Actually you fought twelve (12) days beyond that point with constantly increasing casualties and small loss in combat efficiency.

From the night of 16 November until our relief on 3 December, our Combat Team was the easternmost unit in the entire army and therefore the greatest threat to the German Reich. To stem our attack, units were withdrawn as far south as the Schnee-Eiffel sector and as far north as Merode and flung against us. Enemy forces opposing American units on our right and left were repeatedly withdrawn from those areas and committed against us. Every enemy unit committed against us between 16 November and 3 December was destroyed.

During the last days of this historic battle, many platoons were commanded by private soldiers who had joined us only two or three days earlier. Few companies had more than one officer and often these had just joined us. The fact that Combat Team 22 took every objective assigned it and, the day before its relief, still had the tenacity, the will, and the fighting heart to destroy a fresh enemy battalion, average age 24, and in effectives equal to two of our battalions, is eloquent testimony to the greatness of our Combat Team.

Our infantry, our artillery, our engineers, our tanks, our tank destroyers, our medical aid men, our collecting company, combined in this operation to make the Battle of Hurtgen Forest an eternal part of our country's history and a military classic of all time.

We have finished that battle. We have been moved to this sector to give us an opportunity to reorganize, refit, and retrain. How much time we shall have here no one knows, but we do know that in war, time always presses. Therefore, in justice to our fallen comrades whose faith and courage have carried us this far, we must accomplish our present mission with the same brilliance, tenacity, and aggressiveness that characterized our great victory in the bloody forest of Hurtgen.

There are no words to describe my pride in you or my confidence in you. I can only repeat what has been said to me again and again by those who know your record and who have seen you fight—"You are one of the greatest fighting teams in all American history."

May God keep your courage and faith high, and may He protect and you in the hard battles still before us.

(signed) C. T. LANHAM, Colonel, 22nd Infantry, Commanding.

HEADQUARTERS 22ND INFANTRY
COMMAND POST OF THE REGIMENTAL COMMANDER

A P O #4, US Army,
1 May 45.

In the Battle of Hurtgen Forest during the period 16 November 44 thru 3 December 44, Regimental Combat Team 22 fought and decimated or annihilated the following enemy units, in the order in which they are listed: The basis for this compilation is Combat Team 22 S-2 Periodic Reports numbers 164 thru 180, and Interrogation of Prisoner of War Team 35's reports which accompanied these S-2 Periodic Reports as annexes.

In the Battle of Hurtgen Forest during the period 16 November 44 thru 3 December 44, Regimental Combat Team 22 fought and decimated or annihilated the following enemy units, in the order in which they are listed:

1. 18th G A F Fortress Battalion
2. 275th Feld Ersatz Battalion
3. 20th G A F Fortress Battalion
4. 275th Engineer Battalion
5. 1055th Grenadier Regiment
6. 1031st Security Battalion
7. 1058th Grenadier Regiment
8. Regiment Trier
9. 984th Infantry Regiment

10. 11th G A F Fortress Battalion
11. 1057th Grenadier Regiment
12. 1056th Grenadier Regiment
13. 2nd Panzer Division Reconnaissance Battalion
14. 11423rd Fortress Infantry Battalion
15. Grenadier Werfer Battalion, AOK 7
    (Russion Mortar Battalion) 16...
    .. 191st Feld Ersatz Battalion
    ,, Armee Waffen Schule Battalion
    .. Police Battalion Schamp
    .. 983rd Infantry Regiment
    .. Combat Team Prichmann (Remnants Regiment,
    Company size)
    .. Combat Team Braun (Remnants of 1057th and
    1058th Regiments, and 191st Ersatz Battalion)
    .. 041st Infantry Regiment
    .. 344th Feld Ersatz Battalion
    .. 344th Fusilier Company
    .. 985th Infantry Regiment

The basis for this compilation is Combat Team 22 S-2 Periodic Reports numbers 164 thru 180, and Interrogation of Prisoner of War Team 35's reports which accompanied these S-2 Periodic Reports as annexes.

/s/ Frederick T. Kent, Jr. /t/ FREDERICK T. KENT, JR. Major, 22nd Infantry, S-2.

CERTIFICATE

HEADQUARTERS 22ND INFANTRY
COMMAND POST OF THE REGIMENTAL COMMANDER

A P O #4, US Army, 1 May 1945

The 22nd Infantry Regiment suffered the following casualties during the period 16 November 44 through 3 December 44 in the Battle of Hurtgen Forest.

	Enlisted Men	Officers	
Killed in Action	126	12	
Wounded in Action	1782	77	
Missing in Action	178	6	
Non-battle Casualties	489	8	
TOTAL	2,575	103	2,678

/ s/ John F. Ruggles
/t/ JOHN F. RUGGLES Lt. Colonel.
Twenty-Second Infantry Commanding

13. From Hurtgen Forest To Luxembourg

4 DECEMBER 1944

It was a full day's trip south through Belgium to the Duchy of Luxembourg and the men were very, very tired but relaxed when the convoys pulled in in the dark to the little towns along the Moselle River to take over the patrol of the Moselle. The 83rd Division was relieved that night and the CT was on its own in a strange land where all was quiet; too quiet to the battered squads, platoons and companies that holed up in the little towns like Manternach, Grovenmacher, and Flaxweiler. This was "Chocolate Soldier"—"Prince Rupert of Hentzau"—"Beverly of Graustark" land. It was fantastic and unbelievable. All the people were gone, just as though the Pied Piper had passed this way once again. The houses had REAL beds, the bins were full of apples and potatoes, the deserted cafes were full of wine, a few billy goats and chickens disconsolately poked their way around town—and the Germans were just across the river.

Hurtgen Forest had been a grim tale in many ways. The regiment had taken more casualties in this one battle than many regiments had taken in the entire war. No one, least of all our own command, could logically explain how, day after day, decimated, disorganized, cut to pieces, and under the most adverse conditions the regiment had still fought against over-whelming odds and had still functioned as a fighting machine. As casualties mounted, morale dropped, for as every shell thudded into the earth, there shuddered too into the mind of the disillusioned GI the inescapable fact that the German Army still existed and that the war was far from over.

When at last the regiment had been pulled from Hurtgen and ordered to Luxembourg, the men were stupefied to know that after the heaviest casualties of the war, they were not to be withdrawn from the area and given a rest, but were to be sent down to the Luxembourg front, which even though static, still represented work to be done, patrols to run, and the constant, unceasing vigil which the GI had to maintain under battle conditions.

When the men had moved into Hurtgen, they had gone proudly as members of a great combat team. They had been given the best available benefit of clean clothing and hot food. Now they came out, after 18 days of fighting, unshaven, unbelievably weary, with eyes that were partially vacant. They were really "fed up with the set-up." The sick, sweet odor of death was never far away from any man.

Vehicles were in as bad shape as the men. Jeeps were frequently telescoped in front and rear because while moving in convoy the vehicle had no brakes and no one was able to stop on short notice. It was a haggard, whipped, abysmally unhappy, utterly disillusioned remnant of a great combat team that made its heartbreaking move from Hurtgen, Germany, to Luxembourg.

Not even the warmth of the welcome of Luxembourg nor its undamaged homes and bright modern shops, which might well have been a street from any modern city in America, could interest those GIs. There was still fighting to be done and they had come to the inevitable conclusion that they were the ones who would have to do it.

And yet, not a soldier present, from the high command to the lowest private, so much as dreamed that within a matter of days this same group of men would be forced to face all of the power and might of a great Von Rundstedt offensive whose plan had largely been based upon the German knowledge that the Fourth Infantry Division, as a fighting organization, should, by all rights, have ceased to exist. Indeed, this was so utterly true that on Christmas Eve, 1944, Drew Pearson announced from information picked up from a German broadcast that the Fourth Infantry Division had been completely annihilated.

Early on the morning of **4 December 1944**, Combat Team 22 had begun its motor march into Luxembourg, a distance of 130 miles. The column proceeded via Zweifall, Eupen, Veriers, Pepinster, Theux, Louveigne, Remouchamps, Auwaille, Houffalize, Bastogne, Arlon, Steinfort, and Luxembourg City. Combat Team 22 was to relieve CT331, 83rd Infantry Division, along the Moselle River line by the 5th of December. Thus, each of the rifle companies moved directly to the area held by a similar company of CT331. Special units moved into bivouac areas in the neighborhood of Moutfort.

Shortly after daylight the 5th of December, the actual relief of front line units had all taken place. By dark the operation in the entire area was completed without enemy interference. In its new sector, CT 22 occupied defensive positions on a twenty mile front paralleling the Moselle River in Luxembourg. The usual wide front in the area had been further increased in order to free one regiment to participate in the regiment-by-regiment shuttle switch of the 4th and 83rd Infantry Divisions. When the transfer of the divisions was complete eight days later, the front was reduced by one-third. Three rifle battalions occupied the front with outposts varying in size from half squads (4 to 6 men) to platoons (20 to 40 men). The entire sector was passive and the only aggressive action between forces was occasional harassing artillery or mortar fire and an occasional patrol.

From the 6th to the 15th of December 1944, the regiment continued the outposting and patrolling along the Moselle River and worked to regain the Combat efficiency it had lost. Clean-up of personnel, equipment, and vehicles began at once, A small amount of training was carried on daily, but the Division was in no condition to fight. Nobody expected it to have to fight. The Combat Team frontage was reduced to about thirteen miles on the 13th. When elements of the 8th Infantry relieved the Second Battalion, it allowed that battalion to move to the area, Schrassig Moutfort, as CT reserve. Six to eight weeks were planned to re-habilitate and strengthen the units.

14. The Battle Of The Bulge — Luxembourg

The German offensive struck with lightning speed and in great force early on the morning of the 16th of December, 1944. Initially, enemy pressure was felt north of the Division area and on Combat Team 12. Because enemy pressure was increasing rapidly in the area of CT 12, all units of the Twenty-Second Regiment were alerted for enemy activity and movement plans. The Second Battalion was completely motorized and was to assist the 12th Infantry the following morning. Plans were also instituted whereby the Fourth Reconnaissance Troop would patrol the Third Battalion sector, should it become necessary for that Battalion to move to the assistance of CT 12.

Hitler and his staff knew without a doubt that the Hurtgen Forest Battle had weakened the units which had been engaged. They also knew that the American defensive lines in Luxembourg were thin. German forces were massed for a counter-offensive. The best remaining German strategist, Von Rundstedt, was placed in command. On 16 December 1944, these forces struck the Allied Armies in the Ardennes and started their drive through toward the coast. As the hours passed, it became clear that what Von Rundstedt primarily intended to do was this: In the quick thrust his armies were to penetrate and capture the First Army supply points, demolish communications, and destroy the Americans en route. As rapidly as this was completed, they were to drive northward toward Antwerp, thereby splitting the Allied forces into two commands, a northern and a southern sector. When this was complete, the third successive mission was to defend the southern flank and crush the northern Allied command. This planned area of penetration was into the First Army sector in which the Fourth Division was defending the southern shoulder.

At dawn 17 December, the Second Battalion moved out from its reserve position at Oetrange, some 15 miles northeast, to an assembly area 300 yards south of Bech, Luxembourg. It was now attached to CT 12 to aid in the defense of the Osweiler-Dickweiler sector. Before noon Company F, with Lt. Wilson and Lt. Loyd, en-trucked and moved to Berbourg, where they mounted the tanks of Company A, 19th Tank Battalion of the 9th Armored Division, to go to the aid of Company L, 12th Infantry at Osweiler.

Company F, with the tanks, moved on east into Osweiler, arriving there in the middle of the afternoon. As they neared the town, an American plane attacked the column and knocked out one tank. Identification panels were immediately displayed, and the attack from American planes ceased. By nightfall, the situation was well under control after a very exciting afternoon. Company F remained within the town, while the tanks withdrew to Berbourg.

While Company F had been moving to Osweiler, the rest of the battalion marched on foot northeast from Bech, with companies in column G, E, H. One section of heavy machine guns and a section of mortars were with Company G and a section of machine guns with Company E which left four machine guns, four mortars, 48 men, and four officers with Company H at the tail of the column. The Battalion Command Group was between Companies G and E. At a road junction a thousand yards south of Michelshof, the column turned right on the trail which leads to the eastern part of the woods. About 800 yards beyond the road junction, just beyond the point where the trail crosses the small road to Geyershof, the column encountered a mudhole at least knee deep. They were never able to get any vehicles through this obstacle and from there on the move was entirely by foot.

The advance continued, following the trail through the big woods around to the north along the crest of the ridge. It was just after this move started that the column was attacked. The Germans had evidently been moving south at the foot of the steep bank on the east edge of the woods. They came up this bank at several points, cutting and eventually passing through the American column. The first attack hit the head of Company 'H'. The Germans were apparently as much surprised at the first encounter as were the men of Company 'H'. The first enemy seen was a single scout who reached the top of the bank to find himself face to face with Americans. The German opened fire with a burp gun, but the radio operator of Company 'H' killed him with a carbine.

Then more Germans came up attacking Company 'H' on the right flank and also passing across the front of their column and getting on the left flank. Practically all of the German troops had automatic weapons and they also opened fire with a 50 mm mortar. The men of Company 'H' had been caught flat-footed while carrying the machine guns and mortars. There was a dogfight for a while until Company 'H' succeeded in pulling back a few hundred yards to a draw where they formed a circular defense. There Company 'H' was surrounded for the rest of the afternoon, fighting off an enemy which considerably outnumbered them.

The machine gun section which was at the rear of Company 'E's Column had also been involved in the first enemy attack. Several casualties occurred immediately, including the death of 1st Sgt. Willard of 'E' Company, and as the Germans came up in force between them and the rest of Company 'H', they ran forward to catch up with Company 'E'. It was some minutes before the rest of the battalion knew that Company 'H' was in a fight. Battalion Headquarters heard the firing, but it sounded so distant that they did not suppose it to be in their battalion as they had been hearing considerable remote firing all afternoon. The first information that the battalion had of this attack was when the Executive Officer, Lt. Mason of Company 'E' ran up to the command group and said that Company 'H' was in a fight.

About the same time, the radio operator of Company 'H'—the same man who had killed the first German—got through with a message to the same effect. Col. Kennan ordered Company 'E' under Capt. Faulkner to reverse its direction, move back astride the trail, and relieve 'H' Company. Company 'E' made little progress before they ran into strong German forces and were stopped. The enemy had evidently come up from the east in at least company strength between Companies 'E' and 'H'. At the same time, they came up all around Company 'G' commanded by Capt. Jackson who was almost immediately wounded by mortar fire. For the rest of the afternoon, those two companies were under heavy small arms and 50 mm mortar fire and for a while were separated from each other as well as from Company 'H'. The battalion suffered a number of casualties during the afternoon skirmish which turned into a dog fight on all sides.

That night contact was restored between 'E' and 'G', and they formed an all-around defense circle near the eastern edge of the woods, with the battalion command group and a section of machine guns completing the circle on the north. There was no contact with Company 'H', the last radio message having been received shortly after the attack started when the 'H' Company operator said, "Don't call me anymore; the enemy are too damned close."

Apparently there had been no enemy activity north of the positions of Companies 'E' and 'G' as during the afternoon it had been possible to move to the trail junction about 500 yards north of Company 'G' and back down the other trail to 'E' without encountering the enemy. After dark there was no contact and no firing throughout the night. Evidently the hostile force had gone on its way to the southwest. Company 'H' also was able to withdraw after dark and returned to Bech.

Communication was difficult and uncertain throughout December 17th and 18th. When the battalion advanced from Bech that first morning, a wire vehicle accompanied the command group in the usual way, but it was stopped at the mudhole north of Geyershof. The wire crew then removed the reel from the vehicle and tried to follow the advance on foot, but they were unable to keep up. Throughout the afternoon this crew was by itself in the woods doggedly laying wire, but they never succeeded in reaching the forward CP. Throughout the fight in the afternoon and the passage of the German battalion through the woods, the wire crew was never attacked, but the next day German wire was found tapped in our line. Apparently the enemy deliberately allowed our wiremen to go on with their work.

There was only intermittent wire communication from the CP at Geyershof to the rear. The line back to Bech was cut repeatedly by artillery fire and though repair was continuous, the line was out much of the time. During the night this line stayed in but just south of Bech the wire to the 12th Infantry CP was cut by shelling. Because of this difficulty with its wire, most of the communication to the rear during the late afternoon and night was transmitted through the artillery, which managed to keep in its line from Geyershof to the battery at Bech, which had a line to 42nd FA Battalion and thence to 12th Inf.

There was, however, little positive information to transmit to the Regiment that night. About all that was known at the CP about the situation was that the battalion was surrounded by enemy in unknown strength. There was no wire forward and little use could be made of the radio, partly because of poor reception in the Geyershof hollow and partly because Col. Kennan was unwilling to talk much on the radio for fear of informing the enemy of the situation. Again the artillery battery did an excellent job of maintaining communication in spite of the difficulties. When the artillery radio jeep was stopped at the mudhole, the radio operator showed excellent initiative in finding a way to operate in spite of being left behind. He brought his jeep back to Geyershof and parked it at the window of the switchboard room at the CP. He remained in contact with the forward observer by portable radio, and the next day successfully controlled fire by receiving radio messages from the FO and shouting them in the window at the telephone operator who transmitted them to the battery at Bech. (This man was killed the next day when a shell fragment hit him in his fox hole).

Supply was also a problem. The impossibility of getting vehicles through the mudhole on the route followed by the battalion and the fact that the enemy was putting observed fire on Michelshof, made it necessary to resort to carrying parties. Supply jeeps went as far forward as possible, which was about 200 yards east of Michelshof, and there they were met by the A and P Platoon, which hand-carried everything for the remaining mile and a half to the battalion. It was a source of amazement to the battalion officers that in spite of the enemy battalion, which was somewhere in the woods nearby, the A and P Platoons worked without interference all night and completed the carriage of supplies.

The lone tank destroyer which was in Michelshof was very uneasy about its exposed position and asked for infantry protection. This the battalion was unable to furnish, and the tank destroyer wanted to withdraw. With much persuasion they were induced to remain, Capt. McLean pointed out that members of the carrying party and the jeep drivers would be near their position most of the time. McLean was very anxious to keep the tank destroyer there to furnish a little protection for the sensitive transfer point on the supply line. After the last trip, the carrying party remained with the tank destroyer until morning, which led to their being cut off when the Germans advanced.

On the morning of 18 December, the Battalion up in the woods found itself free of any enemy contact. They moved south to a lateral trail, turned east, and marched toward Osweiler without any opposition from the Germans. But as they came out of the woods on the road, north of Fromburg Farm, they received heavy fire from American tanks, the same tanks that were with Company F. These tanks, which had spent the night at Herbon, had returned to Osweiler in the morning and had been sent eastward to assist the battalion. In view of the situation which had existed the previous night, it was not reasonable to expect that the battalion would march out of the woods unopposed, without the firing of a shot, and the assumption by the tanks that they were the enemy was natural. It was a difficult situation for nearly two hours with the infantry battalion pinned down and suffering several casualties. Eventually a patrol with a white flag made its way around through the draw on the right and made contact with our own tanks.

A little before noon, Companies E and G entered the battered streets of Osweiler and joined Company F. For the rest of that day and the 19th, they remained dug in at Osweiler under very heavy artillery fire and taking no action except the outposting of the town. Here the three rifle companies were reunited, but were far separated from the battalion headquarters at Geyershof and the Service elements and the greater part of Company 'H' at Bech. These elements were ordered to move to Herborn, but before the move could be made, they became heavily involved with the German 316th Regiment.

On 20 December, patrols were sent out to Rodenhof and to the woods where the battalion had been surrounded on the 17th. The latter got well into the woods before meeting enemy and did not return until after dark. The other patrol found the enemy strongly holding the high ground in front of Rodenhof. This patrol got back to the battalion under cover of artillery fire which was directed by the forward observer with them.

The battalion organized an attack on this resistance, getting started in the late afternoon. They got just across the steep ravine which runs southwest from Rodenhof and were stopped by enemy strongly entrenched on the opposite bank. E and F organized their position in the dark for the night with the most forward holes only several yards from the German holes.

The rest of the CT was still down in Luxembourg and because there was no direct contact with the enemy from the 17th to the 20th, Combat Team 22, less the Second Battalion, carried out training whenever possible.

The Third Battalion was relieved of its positions by the Fourth Reconnaissance Troop on the 19th and moved up to reserve positions near Schrassig-Moutfort. The First Battalion was to remain in position attached to CT 8.

* * * * *

Chicago Daily Tribune — December 20, 1944.

1st ARMY LINES SPLIT IN TWO…

Berlin Claims Hodge's men mass to strike at foe's spearheads. Supreme headquarters, Allied Expeditionary Force, Paris, December 19 (AP) The German counteroffensive on the western front assumed the proportions of an attempted major breakthrough tonight.

14 Miles from Capital — a German thrust southwest of Echternach had made some progress beyond Consdorf, which is 14 miles northeast of the little Duchy's capital.

* * * * *

The Combat Team, less the detachments, continued its outposting of the Mosell River line until 1455 hours the 21st of December, at which time they were ordered to move to the vicinity of Bech with all possible speed; the 3rd Battalion arrived at the new area shortly before dark and immediately established a defense of the town.

In the meantime, the 2nd Battalion on December 21st had continued its pressure northeast of Osweiler and made slight gains. They were constantly harassed by enemy attempts to infiltrate, and often the advance completely stopped while mopping-up patrols cleared out the enemy groups behind the lines.

The Combat Team now occupied a sector between CT 8 (on the right) and CT 12 (on the left) with a front of 7,500 yards. Upon arrival in its new zone of action, the Combat Team regained control of the 2nd Battalion defending Osweiler. The 3rd Battalion of CT 12, in a defensive position within the area of CT 22, was placed under control of CT 22. The 1st Battalion still remained in its positions along the Moselle River attached to CT 8. In this location a delaying action was to be initiated.

* * * * *

Chicago Daily Tribune — December 21, 1944.

Supreme Hdq. Allied Expeditionary Force, Paris, December 20 (AP)

The gigantic German counter-offensive is "the big thing" and is increasing steadily in fury.

An Associated Press correspondent said there were three penetrations in addition to the enemy thrust toward Malmedy. A third carried into Luxemburg, south of Echternach, a border town 13 miles southeast of Vianden.

Enemy resistance is also intensifying in the area southeast of Bastogne, in Luxemburg, and from there east to the German frontier where a bitter infantry battle is being fought.

* * * * *

Orders for the Combat Team to counterattack to the north came down on the 22nd of December. The attack was to be made in conjunction with Combat Team 10, 5th Infantry Division (CT 10 was to take over the sector left, now west of the 22nd.) That morning, liaison contact was established with Combat Team 10, and coordination included the 2nd Battalion's helping the northward advance of that team by supporting fires.

H-hour for the attack was set at 1200 hours, but in order to gain more complete co-ordination, this was delayed almost an hour. The 3rd Battalion, the assault unit advanced northeast from Bech, sweeping the area to its objective, the dominating terrain south of Osweiler and Dickweiler. From this position the 3rd Battalion prepared to counterattack to the north.

Elements of Combat Team 12's Battalion were defending Dickweiler, and the 2nd Battalion of CT 22 was, with cooks, clerks, kitchen police, and one platoon of riflemen defending Osweiler. The 3rd Battalion re-enforced the Dickweiler defenses with one rifle platoon and sent Company 'L' to re-enforce Osweiler. The rest of the Combat Team front was covered by platoon strongpoints on dominating terrain, and by the 2nd Battalion's depleted rifle companies on line, facing west of the Combat Ream's left boundary. The rest of that day and the next, the situation remained static with minor skirmishes and patrol action, while the Combat Team waited for CT 10 to pull up abreast on the left.

At 1035 hours the 3rd Battalion was ordered to begin preparations for the relief of the 2nd Battalion, this relief to be effective when CT 10 of the 5th Division came abreast of the 2nd Battalion. The 1st Battalion, still attached to Combat Team 8 along the Moselle River, was relieved and they closed in at assembly areas near Herborn shortly after dark.

CHICAGO DAILY NEWS — December 22, 1944

PUSH THREE-FOURTHS OF WAY ACROSS LUXEM-BURG

Only on the flanks in Luxemburg on the south and in the Malmedy Stavelot sector on the north had the German advance been halted.

GI'S FIGHTING BIGGEST BATTLE, LOSSES HEAVY IN THE STAVELOT SECTOR, BELGIUM

To the south however, elements of five German armored divisions and eight infantry divisions surged forward in new advances with the end not in sight.

CHICAGO DAILY NEWS — December 23, 1944

ALLIED SET BACK WORST SINCE STALINGRAD DAYS

PARIS (UP) — Reiterating a claim that seven American Divisions had been either destroyed or mauled, the German radio named them as the 4th, 28th and 106th Infantry, the 7th, 9th, 10th Tank Division and the 101st Airborne Group.

* * * * *

In the Echternach area, the line has been stabilized. The enemy has been checked in the areas of Dickweiler, Osweiler, and Berdorf. West and south of Echternach in the sector east of Sarregemund we have occupied Uttweiler.

THE STARS AND STRIPES — Saturday, December 23, 1944

FOE 38 MILES FROM FRANCE

On the southern flank where the drive had been halted as early as Tuesday, fighting had stabilized, and all thrusts were checked in the areas of Dickweiler-Osweiler.

"Merry Christmas" was not merry. The men in the rifle companies 'E', 'F', and 'G' up in the woods were exhausted, on two-third rations, no blankets, short on overcoats, no winter boots, short on dry socks, but still long on morale.

On the 25th and 26th of December there was only light artillery fire and some patrol activity. The 1st Battalion relieved the 3rd Battalion of CT 12, thereby occupying the defenses of Dickweiler and positions east of the Combat Team's right boundary. The 3rd Battalion, 22nd Inf. relieved the 2nd Battalion Christmas night.

Excerpts from the diary of Capt. Faulkner, 'E' Company. Note the actions of the first days of the "Bulge" as they affected the men of that unit.

"There is no rest for the weary. We were alerted on Saturday, December 16 to be ready to move in an hour. Got our march order at Bn. CP at midnight from Col. Kenan and made plans to move out at 0700, 17 December.

"We did and joined the motor column at 0730 at the IP. It was ticklish as an air warning was out, the Luftwaffe was active again. By 1100 we de-trucked at Beck, were oriented and moved out at once on foot behind 'G' Company — followed by 'H' — to go to the relief of some 12th Inf. troops in Osweiler and Dickweiler, this side of Echternach. The Krauts had started with patrols and were now crossing the Moselle in force. My Company strength close to 100 with 3 lieutenants, one old timer, Lt. Mason, and two new replacements, Lt. Moore, and Lt. Alderfer. We wondered what the Krauts were up to. No one seemed to know yet for sure. Weather clear and cold with frost.

"Moved single file up the road and east in to the big woods. Things didn't smell right. Too quiet. Moved east into a firebreak trail. Belt of big trees on the right for several hundred yards and the open valley beyond with Osweiler burning in the distance. Very close 2nd growth pine on the left.

"Suddenly, burp gun, MG and M1 fire was heard to our rear. I asked on the SCR 536, "Is that fire on us?" Answer—"It is to our rear." We keep going several minutes until Lt. Mason dashes up. Says our Co. Hdq. and 'H' Company were ambushed and cut off from us. We notified Big 6 (Col. Kenan) at once. Both Companies 'G' and 'E' reversed direction and worked back through the woods to clean out and contact our people. Situation very tense. Wish Lt. Lloyd was with us.

"We contacted people all right, but not ours. Jerry MG fire cut across and into my front, stopping the lead platoon. 'G' Company on my left had a sharp attack. We got about 20 to 30 rounds of mortar which chopped up part of the 1st platoon, wounded the only aid man and almost got me. We got one PW, a wounded aid man, but armed. Sent him to Bn.

"Ordered 3rd Platoon to move right around through small pines to try to envelope the Krauts. 1st Platoon to assist by fire from its forward position, 2nd in reserve, 4th to set up only 60mm mortar we had. Just in the middle of this when S-2 crawled up with orders to dig in at once before it got totally dark. Called back 3rd Platoon, set up defense, and dug in. We had arrived at the ambush point and picked up several wounded from the Hdq. group and a few dead, among whom was 1st. Sgt. Willard. A wonderful man, and I grieved in my heart for him. He had been a 'D' Day man, wounded twice, but always said, "It's not if you'll get it, but when and how bad." He was right.

"It was a weird night. Small arms fire and MG heard in the distance, criss-cross artillery fire coming from all ways, Krauts on our front and flanks, our rear exposed toward Echternach, yet not a round on our position.

"18 December. At first light we dropped in about 12 rounds of 60mm mortar we couldn't carry anymore. Put it where we thought Jerry was. One came out toward us. Nice, well-built, blonde S. O. B. with a MG. We took him, buried the gun, and sent him back to BN. Perhaps he was the one that killed my 1st. Sgt. The men and I wanted to shoot him—but didn't.

"So, we all withdrew back through the small pines in single file. Guess we had Jerry surrounded only he had us surrounded more. 'G' Company led out. Found that Capt. Jackson, just back for the second time since July, had been wounded by mortar fire. Tough luck again. I was only captain with a rifle company. We got news that our 'F' Company had ridden on tanks through artillery fire into the town the evening before and were holding until we got there. Good going for Lts. Wilson and Lloyd.

"The column stopped as we heard MG fire ahead. Found that two of our tanks in the valley saw the head of our column start down from the woods and fired. They wounded several of 'G' Company before Lt. Greenlee ran down waving his maps and stopped them. Fate was being extra cruel.

"I really prayed and asked the Lord to get us the two miles across those turnip fields and open valley and into Osweiler without fire. He did. It might have been at the expense of several of our tanks who were moving southwest out of the valley, as we could see the gray splotches of Jerries artillery shells exploding around them as they passed out of sight over the hill. We went on through the mud and into town.

"Picked our houses in town none too quick, as some big stuff—real big—smashed into the center of town, causing the remaining tanks to keep shuffling around. Jerry had a good OP somewhere. Got a direct hit on the house into which my 2nd Platoon had just disappeared. Sent a runner who reported back they were shaken, but all O. K. Thank the Lord.

"We rested and ate K rations. Went to Bn. CP in school house basement and got new part of town to defend. Back through artillery fire, got everyone on the move around burning buildings, set in position, sent OP out 500 yards to bare hill and settled down for the night. It had been a strenuous December 18th.

"That night and Tuesday were uneventful with spasmodic fire coming in. Sat all Tuesday afternoon in Bn. CP going over three attack plans, each one tougher and more suicidal than the next. Finally told to hold. Returned to a feast at my own CP. Lt. Mason and acting 1st Sgt. Hughes had got all the chickens cached away with six being fried up. Delicious soup, plenty of coffee, homemade jelly, preserved cherries, and a jar of sugared honey found in a closet upstairs. Our thanks to the good house fraus of Osweiler who really knew how to make jelly.

"Worked out route for our 22 man patrol with Lt. Alderfer to take before dawn next day, go five miles, observe, and return by dark. Patrol to be known as "Able Peter". Thus ended December 19th, my 8th wedding anniversary. What a helluva wedding anniversary!

"Wednesday the 20th, saw our "Able Peter" patrol off before first light and we used the rest of the day to perfect defenses, knock holes through stone barn walls for fields of fire, plan AT tank mine and platoon areas. All laid out perfect by 1500 and all changed at 1530 when ordered to move out to northwest at once to the woods, attack east to join 'F' Company who were now moving into the woods opposite Rodenhof, a small village about a mile to the northeast. Find the enemy and attack him! Ugh!

"Got everyone on the way by 1615, except Lt. Mason who could hardly walk with a case of bad swollen trench feet and my "Able Peter" patrol who had not returned. Saw we could hardly get out of the valley before dark, so pushed hard. No time for security patrols. Had to lead with Lt. Moore as scouts. Tough going uphill through muddy fields. To make things worse, two of our tanks pulled up on our right and started shelling the woods to the north to help 'F' Company. Anyone looking toward the tanks could see my men wending their way up. We were tired and weak.

"We saw Jerry as he saw us. He was dug in along a belt of bushes above us—150 yards away. We raced for a wooded draw to his right, got the lead platoon in and opened fire before he did, but his MG caught our 4th Pl. and heavy MG section attacked and cut them up badly. Just dusk now, and we had a real fire fight. Moon with our only LMG got in O. K. and did swell work.

"Our TD artillery forward observer got a battery of artillery on the Krauts. One of our first rounds landed only ten yards to our rear and shook all of us, but then the next volley was on Jerry, and really blew him up. We fired as we dug in. Rather crowded in our little draw and getting dark.

"The town was getting its worst shelling, very heavy artillery and screaming memies (rockets) and burning like a bonfire. We cast long shadows if we stood up. Sent "Frenchie" and another runner to get a guide. They finally returned and we trudged through the mud, in the open, through the woods, uphill and down on the spookiest march I've ever had. Our new tank destroyer F. O. laughed so hard he got hysterics over our adventures of the past three hours, but this was getting to be S. O. P. for us.

"Found 'F' Company in the pitch black dark dug in in deep draws with Jerry only a few yards away. We got into some kind of position after midnight. Tried to dig in the rooty, stony sides of the slopes. Quit at 4 A. M. and laid in the open holes to sleep. Exhausted!

"Able Peter" patrol with Lt. Alderfer arrived at 0725 on 21 December in the dark with orders for us to attack at 0800 with 'G' Company coming out from town. So made a map reconnaissance and issued the order by flashlight under my trench coat. Rough!

"Did move out at 0815. Open draw on our right for 100 yards with wooded ridge beyond. Ridges and deep draws across our direction of attack. 'G' Company moving up on our left at top of ridges and ahead. My plan, 1st. Platoon on right, 3rd on left, 2nd to follow in support, but all hell broke loose again as we moved out. No enemy to our immediate front but heavy mortar fire and all sorts of MG and rifle fire from across the draw on our right flank. No way to get at them there except work around. We kept moving.

"Lt. Moore was shot in both legs. The Sgt. leading the 3rd platoon was lost, and I talked to 3rd platoon radio operator and put him in charge to bring the platoon up the ridges. Sgt. Emberson leading the 1st platoon was shot in arm and did he yell! I moved up to the crest of the next ridge and got blown to the bottom by a mortar round landing on top of the ridge. Radioed word for all platoons to get up to 'G' Company as fast as they could. Ran on along the heads of the ridges. Put command group in a hole and went on with 1st. Sgt. Hughes, Sussman, the 300 radio operator and Frenchie, our runner to get the lay of the land.

"We headed down into the next open draw as a burp gun opened up making dirt bounce at our feet. Jumped high and hit the draw. Found Lt. Greenlee and a few of his command group of 'G' Company there—some wounded. Watch out! Six Krauts coming in counter-attacking over the ridge just over our heads and more following. Only two grenades left in our group, and I had 'em. Tossed one to Sgt. Hughes, we pulled pins and threw them on top of the ridge. Fired the carbine and Lt. Greenlee picked off two with his M1. Nipped that attack in the bud.

"Rest of our platoons came up just as we counted close to 80 or 90 Krauts running across the open into the woods on our left. Got our artillery right on the spot within a few minutes and that stopped that threat and made 'em sweat. The 2nd platoon killed 6 snipers in a draw on our left with a squad under Lt. Alderfer and two on our right.

"Issued orders to dig in at once. We dug deep. Sawed 8" logs to cover holes. Hard freeze that night and the dugouts were like cement pill boxes. Our feet too. Trench foot hit us hard that night with several going out on stretchers. Lt. Alderfer went back with an infected arm, and I was the only officer left.

"Got a lieutenant the next day. Held hard the 22nd and 23rd with enemy holding the same way. Some patrol action.

"On Christmas Eve, one of our month-old replacements crept 40 yards and tossed a hand grenade into the nearest Kraut slit trench. Finis.

"December 25, Christmas Day, our 6 man patrol under Sgt. Mitchell worked 130 yards through the woods to shoot two Krauts in their dugout and steal their MG 34. Swell going for our Christmas present. A light snow and all is quiet.

"Relieved by Company K, starting 2200, 25 December 1944. Sneaked across the frozen fields back to Osweiler, hit the sack for two hours and started a foot march at 0415 back to Herborn and by truck from there to Berbourg. Approximately 50 men and 2 officers now left in Company E. Rest, it's wonderful!"

No one can describe the loneliness of Christmas in combat. The regiment had endured Hurtgen, quite sure that nothing could be worse than this endless slaughter. But Luxembourg and the Battle of the Bulge was worse. The decimated rank of the regiment was forced to cover assigned fronts too large for a full strength division to defend.

The pressure against the regiment by the enemy was so constant as to keep the men at battle pitch for days on end without the necessary time to re-group and to release the tension absolutely vital if men are to endure.

During these vital days, men awoke to fight out of their fox holes; command posts were over run. Mail clerks and cooks joined the battle and fought valiantly for the life of the regiment. Incredibly, it held! The Twenty-Second never broke, even though over run or surrounded. It fought as it had never fought before, and it survived.

Christmas is cold in Luxembourg. The city of Luxembourg was beautiful and modern; the people were uniformly friendly and helpful. But on the front, men were dug into the snow, or existed in ruined buildings, never free from the chilling uncertainty and pressure of enemy attack.

Bits of bright tinfoil or gaily colored paper frequently decorated a foxhole. One insuppressible aid man attached a small Christmas tree to his helmet.

Christmas Eve was bright with a bomber's moon. A communications sergeant cut the switchboard into a radio broadcast and throughout the front, men listened in lonely misery as Dinah Shore sang hauntingly, "I'll Be Seeing You." The music was interrupted to order out a patrol which must cover an extended front to assure the enemy was not trying to move through. Sgt. Heavener led the patrol.

And in a Luxembourg home, Captain Clarence C. Hawkins and Chaplain Boice sat down to dinner as guests. Mother and Father, three children, and Grandmother and Grandfather were their hosts. The dinner, excellent though it was, was second to the friendship and warmth of this family sharing its love with Americans far away from home.

And after dinner, the children watched excitedly while mother and father lighted each candle on a beautiful tree. Then, in the candle shine, the group sang "Silent Night, Holy Night, All is calm, All is bright." Grandmother and Grandfather sang it in German. The children and the parents sang it in French. Captain Hawkins and Chaplain Boice sang it in English with voices choked a bit. In any language, it was understood. In any time or any place it was the hope of men who dreamed of their own homeland, their own fireside and loved ones. And as they went out into the night, only the great candle of God hung in the sky. And the memory of "Silent Night" slipped away as the sound of breaking artillery and machine gun fire made a mockery of "Peace on earth."

Returning to the regiment, Chaplain Boice was called to the aid station. Three of the men on the patrol had been wounded; one had been killed.

It was Christmas.

* * * * *

A letter written from the hospital in England, 6 January 1945.

'Dear Sir: I'm writing you this note in all sincerity, and I pray you will give it your deepest consideration.

'First I want you to know that I don't blame you that you didn't let me come back until we were relieved. I fully understood the situation we were in at the time I asked to be evacuated.

'If those two days had even cost me my feet I feel that Reconnaissance Patrol in which we captured the German machine gun was worth it.

'I'll always feel we were a good outfit, and I would have followed your leadership anywhere.

'Sir, since I had no chance to get my mail I would consider it a great favor and would appreciate it very much if you would personally see that my mail is forwarded to this address as soon as possible.

'God bless you and Company E.'

Sincerely yours,

Sgt. '

CHICAGO DAILY TRIBUNE — 25 December 1944
GERMANS ADMIT YANKS GAIN ON SOUTHERN WING

London, December 24 (AP)

The Berlin radio tonight admitted that advances had been carved into German lines by American forces on the southern flank.

SUNDAY EXPRESS London, December 31, 1944
PATTON HAS ONLY 12 MILES TO GO — TANKS CUTTING RUNDSTEDT'S SALIENT

Elsewhere on this southern flank are massed two other armored divisions, the 9th and 10th and four infantry divisions, the 4th, 5th, 26th and 85th.

CHICAGO SUNDAY TRIBUNE—December 31, 1944.

With the U. S. 3rd Army on the western front, December 30 (AP)

High praise for the 4th infantry division for saving Luxembourg was expressed by Lt. Gen. Patton in a letter today to the divisions commander, Major General Raymond O. Barton of Ada, Okla. "Your fight in the Hurtgen Forest was an epic of stark infantry combat" said Patton, "but in my opinion, your most recent fight—from December 16 to December 26—when a tired division you halted the left shoulder of the German thrust into American lines and saved the City of Luxembourg, is the most outstanding accomplishment of yourself and your division."

The 12th Corps commander, Maj. Gen. Manton S. Eddy, also sent a letter of commendation to Barton, noting the division's record since D-day. "There are other divisions of the 3rd Army which can report the same sort of story of sudden moves, hungry mornings, cold and hard fighting—all of which spelled disaster for the German breakthrough attempt on the southern flank.

* * * * *

The 5th "Red Diamond" Infantry division was fighting in the Saarlautern bridgehead when the order came for it to switch to the north.

Twenty-four hours later, the 5th had covered 70 miles to take up reserve positions north of Luxembourg City.

Forty-eight hours later a combat team of the 5th launched an attack on Reinstedt's southern flank at Echternach and within another twelve hours the whole division was passing through the 4th Infantry Division which had borne the first shock of the Nazi offensive and was attacking as a unit.

The Combat Team's boundary was shifted almost 2,500 yards west on the 27th of December. In view of the change, the 3rd Battalion moved two rifle companies west, swept the intervening areas, and relieved elements of CT 10 at 1800 hours. The 1st Battalion moved Company 'C' to Osweiler and relieved the elements of the 3rd Battalion defending the town.

During the day, the 1st and 3rd Battalions established outposts to the north and northeast and sent vigorous patrols probing to the front. These patrols were able to reach the Sauer River and Echternach, which were enemy-occupied. The 2nd Battalion remained in mobile reserve in Berbourg.

While the Combat Team was defensively deployed near Echternach, Luxembourg, a very amusing incident occurred to several men of Company 'M'.

Overlooking Echternach from the south was an extremely high hill. Situated on the back slope of this hill was S/Sgt. Miller and his section of 81mm mortars. From the top of the hill and from the side of the western base, it was possible to see the German fortifications and often slight movement. In this section was a rather new replacement, Pfc. Bryan, who had previously served with the 2nd Division. This man loved excitement in its most dangerous form. Sgt. Miller and Pfc. Bryan, while lying in their foxhole one cold snowy night requested permission of their platoon leader to go down to the river the following morning and fire at the Heinies. Shortly after dawn these two ambitious soldiers struck out across the snow. Behind them lay their wind-swept trail and the other men of the section, who were wondering if they would ever see these two again. This task was extremely hazardous, for these men were under enemy observation and fire for almost eight hundred yards. Taking advantage of all possible cover, they maneuvered into the first line of houses along the river bank.

Immediately they set their radio into operation and reported their position in code: "Tare-Item to Tare-Fox, Tare-Item to Tare-Fox, Over." T/Sgt. Romeo Bolduc answered from the gun positions, "Tare-Fox to Tare-Item, Tare-Fox to Tare-Item, Send your message. Over."

The reception faded and became almost indistinct, but after several anxious minutes of waiting, the message came through. "Tare-Item to Tare-Fox. Donkey men OK. We are going on to our position. We are now at Able seven-niner, Sugar one five. Over."

"Tare-Fox to Tare-Item. Roger. Out."

The two men had found in the first house a phonograph and several records. These they started playing, and they sampled the latest cognac before continuing.

After twenty or thirty minutes, the two men slipped from the house and went along a stone wall which paralleled the river and provided a means of partial defilade. When the wall terminated, they dashed across the open space to a railroad embankment which ran along the river bank. Proceeding south almost 200 yards, they located a culvert under the tracks and decided the CP was to be here.

Pfc. Bryan secured the radio and called the platoon CP: "Tare-Item to Tare-Fox, Tare-Item to Tare-Fox. Fire concentration six. Over."

Back at the command post the radio hummed and the message broke through. Gun crews quickly laid on the target, and in a few minutes the dull, unmistakable cough of the mortar was heard. Then the message, "Tare-Fox to Tare-Item, On the Way. Over."

As Sgt. Miller and Pfc. Bryan tensely awaited the explosion, they looked for another target. Suddenly black smoke drifted up 100 yards in front of them. That was perfect. Now to find a Kraut and let him have it. While they waited for enemy movement, they fixed numerous concentrations on located positions. After several hours, Sgt. Miller tapped Pfc. Bryan on the shoulder and pointed to one lonely German walking along the far edge of the river. They had to restrict their talking to mere whispers, as it was quite easy to hear several Germans talking in dugouts across the river.

Pfc. Bryan grabbed the radio and said, "Let's get that bastard before he knows what's up." The radio buzzed, and the fire mission was delivered to the gun crews. All six guns laid on the target and, at a command from the platoon, fired one round. In a matter of seconds, the rounds hit, completely circling the German, wounding him slightly, for he took off like a streak of lightning for the nearest bunker.

Back at the CP, everyone waited nervously for the adjustment, but none came. Capt. Offt, the radio operator, tried frantically to contact the two observers, but it proved impossible. Everyone tried to act unconcerned; then Sgt. "Juicy" Walters broke the silence saying, "They are probably just moving to a better CP and turned the radio off." The men tried hard to believe this, but as darkness fell and the radio still had not made a noise, the hopes for the observers' returning was about gone.

As soon, as it was totally dark, when the waiting had become unbearable, a patrol was organized to search the river line for the two men. Just as the patrol moved out, the message came in to Pfc. Bowen, the radio operator with the patrol, "Donkey men returned to Charlie Peter. Everything O.K." The patrol returned to the platoon CP.

Later, in talking to the two observers, it was learned that they had purposely turned off the radio because the enemy had spotted them. Later, when they tried to re-establish contact, the aerial broke and the radio was put out of order. Sgt. Miller related the story that our own artillery shells had fallen both in front of and behind them as they lay in the culvert. As they left the CP, they said, "Well, Lieutenant, if you want us to do it again just say so and we're ready. It was a lot of fun, and boy, did we get a good phonograph."

From the 28th through the 31st of December, Combat Team 22 maintained and improved its screening of the Sauer River. Additional observation and listening posts were established and coordinated; reconnaissance and combat patrols were maintained; and defensive positions in depth were improved. Several times throughout this period, under cover of darkness, patrols crossed the river to locate enemy positions and returned safely.

Combat Team 22 was defensively employed during the period, 1 January 1945 to 13 January 1945, and held part of the southern shoulder of the bulge created by the German counter-offensive. It was responsible for the Osweiler-Dickweiler sector in Luxembourg extending along the south bank of the Sauer River.

While the First Battalion was located in Osweiler, Lt. Col. George Goforth moved his forward CP to a village almost on the banks of the river, and almost certainly under the observation of the enemy. There were only three houses grouped together and in the basement of these three houses the CP took its stand. The Colonel, with his S3 and S2, were in one basement. Lt. George Kozmetzsky had the forward aid station in the second, and Capt.

Reece (Daddy) Dampf, adjutant of the First Battalion, had his Communications and I. and R. Section in the third basement.

Every effort was made to keep any activity from being observed by the enemy, when along toward dusk, high in the lean-to next to Daddy Dampf's CP, in some mysterious manner, probably by a careless cigarette or a G.I. heating coffee, a fire started. The consternation caused by the conflagration was great, especially in the face of approaching darkness and the fact that the burning buildings lighted the entire CP like a signal fire for miles around. The efforts of Daddy and his men to put the fire out were totally ineffective, but increasingly amusing to those who were observing from a distance. Not only was Daddy Dampf warm from the fire and from his personal exertion but was made considerably warmer by the caustic comments from Col. Goforth, whose vocabulary was certainly not limited. Thus, the CP of the First Battalion was moved, and it was not due to enemy action.

The greater part of the enemy had been forced to withdraw to the confines of the Siegfried Line fortifications on the northern side of the Sauer. Combat activity was confined to harassing and interdictory artillery and machine gun fire and patrols, combat, reconnaissance, security, and contact.

Because the enemy was able to cross the river during the hours of darkness and establish listening and observation posts, there were numerous patrol clashes and several ambushes. This could be done because of the wide frontage and the wooded, rugged terrain, but the Regiment worked continuously to improve its positions. The defense was prepared in depth, fires were better coordinated, protective wire was strung, road and bridge demolitions were prepared, mine fields were laid, and booby-traps set.

Whenever possible, training was carried on. Emphasis was

placed on such topics as flame throwers, demolitions, mortar drill, and small unit assault training, to include the use of the assault beat. Reorganization and indoctrination of re-enforcements continued.

On the 13th of January, plans were formulated, and orders were issued for the 2nd Battalion to clean out the tongue of land formed by a bend in the Sauer River some 2,000 yards below Echternach. A deep wire entanglement, covered by automatic fire, protected the base of this peninsula, and further re-enforced the defenses of a strong enemy outpost believed to occupy the tongue. Company 'G' was assigned the mission.

By 0400 hours the following morning, January 14th, the Scouts, and Raiders of the 2nd Battalion had blown diversionary gaps in the enemy protective wires. As planned, Co. 'G' further gapped the wire by using bangalore torpedoes, and one platoon passed through the gap, turned southeast, overran several enemy bunkers, and swept almost all the peninsula. The mission complete, Co. 'G' returned to its former defensive positions.

Information was received the next day that CT 22 was to be relieved by Combat Team 347, 87th Infantry Division, then to move to the 5th Division area some 12 miles west and relieve Combat Team 11. Staff planning and coordination began at once. Liaison contact between relieving units had been established by dark and final preparations were made.

Combat Team 22 was relieved by Combat Team 347 on the 16th of January. The 3rd Battalion, first relieved, moved motorized from its reserve location in Berbourg to the village of Haller and there relieved the reserve battalion of CT 11. The relief of the 1st and 2nd Battalions, their move, and their assembly in the town of Junglinster were delayed. Trucks were late, roads were poor, and the forward positions received a heavy shelling during the relief. The relief was completed at 2200 hours, and CT 347 assumed the responsibility for the front.

The 1st and 2nd Battalions assembled in Junglinster, closing in shortly after midnight. At 0830 hours the complete relief of CT 11 was affected, and by 2100 hours that night it was complete. Because several outposts were forward and could not be relieved until after dark, the delay was necessary. The CT occupied the same defensive positions as had been held by CT 11.

During the week of the 18th to the 25th of January, the Combat Team maintained defensive positions similar to those previously held at Osweiler. As before, the defenses were along the Sauer River line, with the front lines extending along the western banks of the river. The type of terrain in this location was more favorable for a defensive position, The front extended from the junction of the Sure and the Our Rivers, which formed the Sauer, east approximately 800 yards. The river line was screened with outposts during the day and listening and contact patrols at night. The defense was so organized that there were two battalions on line occupying strong points on dominating terrain. One battalion remained in reserve ready at any time to move into position and support either of the forward units.

Throughout the period the defense was improved. Demolition sites were prepared, wire and mine fields were laid, booby-traps were set, and 81mm mortar platoons fired harassing fire on targets of opportunity or possible enemy emplacements. Delaying positions to the rear of the main line of resistance were constructed and coordinated. Reorganization, as before, was carried on in conjunction with unit training.

Information was received the 26th of January 1945, of a pending relief of the Regiment. The CT was to be relieved by elements of the 80th Infantry Division and then was to assemble near Trois Vierges.

CT 22's relief began during the afternoon of the 28th by CT 318, 80th Division. The 1st Battalion, upon relief, moved to Medernoch, the 2nd Battalion, moved to Waldbillig, and the 3rd Battalion moved to Christnach.

The Combat Team was completely assembled and prepared for the motor move to its new sector by 0330 hours the following morning. In compliance with orders, the leading elements of the combat Team crossed the IP at Laroche at 0735 hours. The movement was made via Sauel, Reichlange, Attert, Bastogne, Trois Vierges, to a destination near Huldange. By dark, the move of seventy miles had been successfully completed despite ice, snow, and extremely heavy traffic.

Following the decisive battle of the Fourth Infantry Division for Echternach in which the Second Battalion played so vital a part, the Regiment assumed a defensive position on the Mosel River. With the First Battalion Headquarters in Gostingen, the Second Battalion Headquarters were located in Flaxweiler. It was during this period of time that several officers previously wounded and hospitalized in England were returned to duty with us, and thus came to light one of the strangest stories of the war.

On D-Day, Captain Tom Neely, S-3 of the Second Battalion, had landed with the Third Battalion previously in order to be abreast of the landing situation and in order to arrange proper planning and attack of the Second Battalion as it came ashore. This Captain Neely had done in his usual efficient way, and he was the first person the Second Battalion met when they hit the beach.

It was at the close of the day that Captain Neely and Major Earl W. Edwards, the Battalion Commander, were walking down a trail together in the vicinity of St. Martin. It was not yet completely dark but was dusk enough that Major Edwards and Captain Neely had to use a flashlight to read the map which they held as they were checking positions. They had just finished the map check and had turned again down the trail when one short burst of machine gun fire cut the air and Captain Neely dropped, shot squarely through the stomach.

The burst of machine gun fire had come from an American machine gun manned by a paratrooper. On checking the story, it seemed that the paratrooper had been stepping over his gun position and had apparently kicked it, causing it to fire one burst, the burst which hit Captain Neely.

There was no man in the regiment more beloved than red-haired, congenial Tom Neely. He was a fine officer and a Christian gentleman. Major Edwards immediately called for the battalion surgeon and Captain Neely was placed on a litter and carried into a farmhouse in which lay other wounded, and next to a building which was burning brightly. The surgeon could do little but administer morphine in an attempt to alleviate the pain. No ambulance could get up the road, part of which still lay in enemy hands, and even so, there would have been no place to evacuate him since the surgical teams and the hospitals had not yet come ashore. There was nothing to do but to stand by the side of this man whom we knew and loved and to do for him the best we could.

It was here perhaps that the men of the Second Battalion realized with a shock that war was hell, that there was nothing glorious or wonderful about it, that it killed and maimed, that it hurt and terrified, that it was sordid and ugly. We remembered a wife in Carolina and a little boy named Dan with soft brown eyes and a stubborn chin like his dad's which he had a habit of pulling down when he talked in a gesture so much like Tom's that most of us called him "Tom, Jr."

Throughout the long hours of the night we watched the grey pallor of death as it came closer and closer, and wondered a thousand times the answer to the question which Tom constantly put to us, "Why did this have to happen; why did it have to be?" About 3:45 a.m. Tom died. We arose then from our watch and because there was nothing further we could do, we went on again to the troops. Little did we realize that death soon would become our bedfellow and there would be little time to spend with those who were dying, but rather with those who were attempting to remain alive.

And so the strange end of this story came to us in Luxembourg when Captain J. O. Jackson, a good friend of Tom Neely's, and wounded in the Corridor of Death at Periers, returned to us. He told us that when he had passed through the Tenth Replacement Depot he had been in the hospital, and passing the psychiatric ward when a soldier called out to him and said, "Captain, are you from the Fourth Division?" "Yes," Jackson replied. "Do you know anybody in the Twenty-Second Infantry?" said Jackson.

The soldier's eyes clouded, and he said, "Did you know a red-headed captain by the name of Neely?" Jackson looked straight at the boy and said, "Tom Neely was one of the best friends I had." The boy's eyes dimmed, and he turned away as he said, "I am the one that shot him."

When Jackson talked to the surgeon in charge of the boy's ward, the surgeon said that the boy was so troubled mentally and apparently so affected physically that he did not believe the boy could ever be all right, nor did he think he would live long.

An understanding and Christian family back home sent immediate word that they held nothing against the paratrooper and asked that he be told they were sure it was an accident and that they forgave him, but when the word arrived no one knew who he was or where he was, and nothing could be done. It was another one of the vagaries of war.

15. Repeat Performance — Second Penetration Of The Siegfried Line

The great German counter-offensive led by Von Rundstedt had been smashed in the past month of action. In reality, the Ardennes offensive had ended about the middle of January. Since then, the American forces had been reorganizing their lines and preparing to push the Germans back across the Rhine River. During the German drive, the enemy was able to again secure their Siegfried defenses. The Germans were now sending troops into the line and preparing to defend the Homeland. Because the Allied Armies had previously penetrated the Siegfried Line, the Germans found the deficiencies in the structure of the Line and were making every effort possible to correct these.

Even though there was snow on the ground on the 1st of February, spring was in the air. Days were longer and the sun shone brightly. Our aircraft were seen daily leaving their misty trails through the clear skies as they streaked for the German interior. It was a glorious sight; thousands of huge bombers, supported by fighter planes roared continuously; bomb-run formations developed before one's eyes; and enemy ack-ack fire specked the sky. As the Combat Team moved eastward along the highways, it was possible to see the destroyed German vehicles, horses, and trains that the Air Corps had previously wiped out.

The defeat of the counter-offensive brought with it initiative, renewed courage, and a surge of power. The troops now realized that they could crush the enemy in spite of what he planned. On the eastern front, the mighty Russian war machine had begun to plow across the captured countries. It was becoming more apparent each day that Hitler had exerted his main potential reserves in the Ardennes strike and had suffered severely in this great loss.

On February 1, 1945, Combat Team 22 was assembled near Hautbellain, Luxembourg, Combat Teams 8 and 12 were pushing on toward the Siegfried defenses near Bleialf to secure a jump-off line for an assault to be made on the Siegfried Line. Minor adjustments in assembly areas were made during the day; the 1st Battalion staged forward to Oudler, Belgium; the 3rd Battalion shifted its position slightly to within the town of Huldange; and the 2nd Battalion remained near the town of Hautebellain, For the remainder of this day and the next, all units of the Regiment remained in place.

The 44th Field Artillery was hurriedly repairing their gun tubes and mounts for the oncoming attack. They had fired well over 100,000 rounds by this time and the artillery wanted to give the best possible support to the infantrymen. The forward observers from the 44th were relaxing a bit from their strenuous jobs. In the field artillery, it is the forward observer and his radio party who have the front line duty to accomplish, for they are continuously with the forward-most elements of the rifle companies so that they can direct effective artillery fire whenever called upon. Oftentimes the field artillery radios were the only sets in operation and all vital communications were transmitted through them.

During the night of 2-3 February, reconnaissance parties

checked the routes forward to new locations. The Combat Team moved out that day at 0745 hours in the order 1st Battalion, 3rd Battalion, Regimental Headquarters, and then the 2nd Battalion. The Regiment moved via Hautbellain, Oudler, St. Vith, Schonberg to forward assembly areas in the proximity of Schweiler, Germany.

Attacking elements of the Division (Combat Teams 8 and 12) made excellent progress, and Bleialf was captured that afternoon. In view of this, one company of the 1st Battalion, 22nd Regiment, moved eastward to secure Buchet, the proposed jump-off point for the Combat Team's attack on the Siegfried Line. By dark, word was received that patrols of the 8th Regiment had entered Buchet uncontested. The entire 1st Battalion, therefore, moved into Buchet before midnight. Combat Team 22's leading elements were now within a thousand yards of the first belt of Siegfried Line pillboxes. That night a strong 1st Battalion patrol moved out to investigate the Line.

Company C, 4th Engineer Battalion, Company C, 70th Tank Battalion (medium tanks), and Company B, 610th Tank Destroyer (90mm. self-propelled) Battalion were all attached to the Regiment at 1500 hours.

The decision was made, in planning the attack, to again assault the Siegfried Line from Buchet. From the eastern edge of town, covered approaches led to within three hundred yards of the woods concealing the first Siegfried fortifications. This route of approach was north of the regimental boundary, but, upon request, Division provided a temporary north boundary to include this approach. The scheme of maneuver was almost identical to that used over the same ground the 14th of September 1944; an attack from Buchet in a column of battalions.

As D-Day was set for **February 5, 1945** the CT planned to

use February 4th to complete last minute preparations, assemble the battalions well forward, open roads for the supporting armor, organize field artillery positions, and reconnoiter the Siegfried Line and its approaches. However, at 0003 hours information was received that CT 8 (north of CT 22) would continue their attack and push on into the Line before dawn; and, in addition to this, a First Battalion patrol from CT 22 reported that they had penetrated the Ormont-Brandscheid Road and had found the first line of bunkers unoccupied, the second line unoccupied. In order to take advantage of these unexpected developments, Col. Lanham, at 0505 hours, ordered the First Battalion to attack at dawn.

Little time remained to alert the troops. At once, company officers awakened their commands and briefed the troops on the situation. The men arose from their wet foxholes in the misty fog to quickly eat a K ration and then meticulously checked their weapons. Those men who had been here before knew the Siegfried Line was going to be a tough fight. This time they had orders not to penetrate the Line, but to breach it and drive to the open plains that stretched west of the Rhine.

The 44th Field Artillery Battalion hastily alerted the remaining men of the gun crews and opened additional ammunition crates. Forward observers with the infantry companies hurriedly organized their parties and made preparations to move forward.

Then as the first rays of light broke over the wooded horizon came the command, "Let's go, men; this is it!" The First Battalion moved out in a column of companies proceeding along the corduroy road that led to the first line of bunkers. Contrary to the previous reports, all pillboxes were found occupied and defended, but resistance varied from moderate to light. Just before the assault on the first pillbox, the old veterans who had been here before saw, still hanging in a tree, the trousers of a man who had stepped on a mine in September and had been blown to pieces. It was an eerie feeling; this sudden remembrance of a comrade who in a matter of a split second had been annihilated. This might happen to anyone anytime now. The Germans knew the war game and meant business.

By noon this battalion had pierced the first line of fortifications and had cut the main northeast-southwest road into Brandscheid. One rifle company was in position forward with the other two companies echeloned to the right and left rear.

The 3rd Battalion assembled in Buchet and followed directly behind the 1st. As the column of men moved through Buchet, they were forced to step over a pool of blood and pieces of flesh left behind where a man had been sliced to ribbons by flying shell fragments. The Jerries were well zeroed in on Buchet and could lay down artillery and mortar barrages any place in the town. The 81mm mortar platoons, trying to dig in within Buchet, were continuously subjected to artillery fire and suffered heavily in casualties.

By the middle of the afternoon, the 3rd Battalion had passed through the 1st Battalion, turned south, and continued the attack along the main road. After severe fighting through the heavy woods, the battalion, without benefit of armored support, advanced and gained control of the crossroads at Meisert. It was now quite apparent that the Germans in their brief offensive had begun improvements on the Siegfried Line. There were to be found connecting trenches in front of each fortification, added camouflage, repaired embrasures, and foxholes intermittently spaced about each of the bunkers.

The 1st and 3rd Battalions dug in and secured for the night; the 1st Battalion outposted the ridgeline just east of the Ormont-Brandscheid Road; the 3rd Battalion remained at the

Meisert crossroads. The initial penetration of the Siegfried Line had been successful, and casualties were comparatively light.

Before the 3rd Battalion could successfully capture Brandscheid the next day, the roads had to be repaired and supporting armor brought forward to the assault companies. That night the entire 2nd Battalion and the 4th Engineers worked to clear the road. Colonel Kenan and Captain Burnside, the Battalion Commander and Executive Officer respectively, worked knee-deep in the half-frozen mud alongside the men and officers in an effort to get the armor through. It had to be done and quickly. Before dawn February 5th, the road was clear.

Four hours later, the 3rd Battalion attacked Brandscheid with two platoons of 90mm self-propelled tank destroyers, and two medium tank platoons attached. Capt. Lee led 'I' Company and Capt. Roche led 'K' in the assault. A dense fog had settled over the landscape, and confusion arose. In the midst of the confusion, the two companies were able to secure their assigned pillboxes. Col. Lanham came down shortly after the attack got underway in order to observe the operation. Capt. Roche said that he didn't have time to explain to the Colonel what was going on. The Colonel must have sensed this, as he remarked, "Captain, don't let me stop you."

As the Battalion moved across the crossroads at Meisert, fighting became intense and eleven supporting, protecting, pillboxes were nullified. By the middle of the afternoon, the strongly fortified town of Brandscheid had been taken, and mopping-up operations continued.

Before dark, Lt. Perkins, with his platoon from Co. 'K', moved to the left flank of the company sector to occupy several pillboxes in that area. By dark, he had deployed his squads into the vacated bunkers and his defense for the night was organized, Lt. Perkins and his platoon were in a very precarious position, as was learned the following day.

In the meantime, the 1st Battalion echeloned to the left rear and followed the attack of the 3rd Battalion. In order to establish a defensive left flank, the 1st Battalion moved approximately 2,000 yards to the southeast along the Bleialf-Sellerich Road. The 2nd Battalion moved from its reserve location in Buchet to occupy the 1st Battalion's former positions.

Late that day, CT 22's zone of action was shifted to the north, and the directions of attack shifted southeast toward Prum. Brandscheid was now in the sector of the 90th Infantry Division. Preparations were made immediately for the relief of the 3rd Battalion by elements of the 90th Division.

The relief of the 3rd Battalion in Brandscheid began at 0445 hours on February 6. As the relief was being made, a strong enemy counterattack, estimated to be five hundred men, hit Brandscheid from the southeast. Co, 'L' which had been completely relieved, immediately counterattacked. A close-in bitter struggle ensued as the enemy, in the confusion of the relief and the blackness of the night, infiltrated throughout the town. Casualties mounted, and the relief was becoming more confusing by the minute. The only usable road into Brandscheid was a mass of vehicles—some moving forward, some to the rear. By 0900 the situation was stabilizing. Lt. Perkins and his platoon had been completely surrounded and were fighting to drive the Germans back. By the middle of the afternoon, the relief was complete, with the exception of the one platoon.

At 0845 hours, in spite of the situation in Brandscheid, the 1st and 2nd Battalions of CT 22 attacked abreast, 2nd Battalion on the left, along the axis of the Bleialf-Sellerich Road. The 1st Battalion captured Sellerich, Herscheid, and the high ground just east of the two towns by late afternoon. The 2nd Battalion met light resistance and by dark had taken Hontheim and the high ground just east thereof. The 3rd Battalion tied in to the right rear of the 1st.

Having taken the high ground east of Herscheid, Sellerich, and Hontheim, the Combat Team again attacked toward Prum on the 7th. The attack was to move through the three towns of Obermehlen, Niedermehlen, and Steinmehlen. In these three towns, one of the war's hardest localized battles was to be fought.

A platoon size combat patrol from Company 'E' moved out before daylight to seize the dominating high ground west of and between Niedermehlen and Obermehlen, called Objective 11. The remainder of the company followed, and the objective was secure by 0830 hours. Co. 'G' bypassed Objective 11 to the north and, after eliminating machine gun and small arms resistance, took Obermehlen. Further advance by the 2nd Battalion was prevented by enemy counterattacks. Co. 'E', on Objective 11, was hit by an infantry tank attack from the direction of Niedermehlen and was forced to withdraw from the hill. A similar force hit Company 'G' in Obermehlen, but no ground was lost. Shortly before dark, Co. 'F' attacked through Company 'E' and recaptured Objective 11, from which they could re-enforce the fires of Co. 'G' in Ober-mehlen.

At 1000 hours, the 1st Battalion moved out to seize the ridge line approximately 1,000 yards west of Niedermehlen, Objective 14. Favorable progress was made, and the ridge was secured by 1430 hours. At once preparations were made to attack Steinmehlen to the southwest. An enemy counterattack, even though repulsed, stopped any further advance of the 1st Battalion.

Because of the heavy enemy pressure on the 2nd Battalion,

the 3rd Battalion moved two rifle companies behind the 2nd to secure and to serve as a counterattacking unit.

After dark that night, Lt. Perkins, who with his platoon was still isolated, sent out a two-man patrol to find out who held the town of Brandscheid. This patrol after many narrow escapes, found a CP of the 90th Division and returned to the platoon with this information. Before dawn the next day, the platoon escaped back through the lines and later joined their company near Sellerich.

The terrain surrounding Prum, Germany, consisted of huge rolling hills which afforded long distance observation of troop movements. From the high ground just east of Sellerich, it was possible to look down into Niedermehlen and up to the hills surrounding Prum. All roads running into this area could be observed by the enemy for miles and as a result vehicle travel was greatly restricted. A vehicle driver in this type of terrain had to be a calm, collected individual. Oftentimes drivers were expected to travel alone to secure ammunition or rations and to return to the front lines. When sitting in a vehicle with the motor running, it is impossible to hear the deathly whistle of incoming artillery; mortar barrages creep down like a swarm of bees; direct fire weapons find no better target; and snipers easily pick off the occupants. There are no safe jobs in the infantry, only varying degrees of comparative safety.

On the 8th of February, the objective, the German communication center of Prum, still had not been taken. The 3rd Battalion was to relieve the 2nd Bn. on Objective 11, seize Niedermehlen, and be prepared to continue the attack to the high ground about three hundred yards west of Prum, known as Objective 8. The 2nd Battalion was to push out from Obermehlen and seize the high ground east of Niedermehlen, Objective 15. The 1st Battalion, from its position on the ridge line overlooking Niedermehlen, was to support the attack by fire.

Due to constant enemy counterattacks from the town of Niedermehlen, the 2nd and 3rd Battalions were not able to make an advance. The 1st Battalion was relieved of its objective, Steinmehlen, by elements of the 12th Regiment and reverted to regimental reserve in the rear of the 2nd Battalion.

The plan for the 9th of February was different, in that the 1st Battalion was to pass through the 2nd Battalion and seize Objective 15, leaving the 2nd Battalion in Obermehlen.

In support of the attack, the 2nd Battalion built three foot-bridges across the stream just east of Obermehlen. At 0850 hours, the 1st Battalion jumped off, passed through the 2nd Battalion, and leading elements crossed the stream within an hour. Companies 'A' and 'C' reached and secured the crest of the hill, Objective 15, by 1030 hours, but very effective enemy machine gun fire from both the north and south prevented Company 'B' from crossing. Five tanks and over two infantry companies counterattacked the hill but were driven back with severe losses.

The 3rd Battalion, by means of patrols, continued to probe Niedermehlen, but met very little success. The situation remained critical until late evening, then after a heavy artillery preparation, a coordinated attack was launched on Niedermehlen by the 2nd and 3rd Battalions.

The 2nd Battalion attacked from the north down the road leading from Obermehlen to Niedermehlen. It was a flat level route check-marked by shell holes, and the turf was dug up from flying bits of shell. The 3rd Battalion, after probing for a secure route, finally drove in from the southwest in conjunction with the 2nd Battalion. Within two hours the town was secured. The captured commander of a battalion of the German 2nd Panzer

Division stated that his entire battalion was lost in defending Niedermehlen.

Resumption of the attack on Prum was planned for the 10th of February. The 1st Battalion was to move from its position to Objective 15 and seize Objective 17, a wooded hill overlooking Prum from the north. The 3rd Battalion was to push from Niedermehlen, up the long hill overlooking Prum from the west and seize the ridge line, Objective 8.

Shortly after daylight, the attack was underway, with the 3rd Battalion leading. This battalion had to push up a long, gentle, sloping hill to Objective 8. From Niedermehlen it was possible to observe each man as he moved forward. It was like the combat films made in training with squads advancing in skirmish lines, by squad rushes, or individually. The attached machine gun platoons would displace forward from one covered position to another, firing continuously. Some gunners even ignored the basic tripod and fired the machine gun from the hip. 81mm mortars fired heavy concentrations of smoke in order to screen the advance. Opposition was unexpectedly light, and by noon the 3rd Battalion was on the objective.

The 1st Battalion met heavier resistance and was able to reach the five point crossroads at Tafel only after bitter fighting. The 2nd Battalion moved on to Objective 15 and took over the positions previously held by the 1st Battalion. During the attack Capt. Newcomb, S-3 of the 2nd Battalion but now leading 'E' Company, was killed. He was one of the best officers and leaders.

Late that evening, orders were received by the Combat Team Commander to take up a defensive position. In accordance, the Commander organized a defense of the sector. The main line of resistance was to include the ridge directly west of Prum, extending north to the crossroads at Tafel, then northeast to tie in with

CT 8, approximately 1,000 yards west of Hermespand. Two battalions were to be on line, and one battalion was to be in reserve in Gondenbrett.

During the afternoon, Captain Lee of Company I and his mortar observer party were in the lone house atop the Prum ridge looking into the town itself. Suddenly and without warning, several 88's exploded in the house beside them. This caused serious damage to Company I and the command group. Without hesitation, Sgt. Le Flamme of the 81mm mortar platoon went forward to seek out the German guns and destroy them. After several hours observation, he located one of the guns and destroyed it with mortar and artillery fire.

Activity during the night was confined to intermittent artillery and mortar fire. With the light of dawn, the Third Battalion reported only negligible enemy activity within Prum and on the high ground to the north. Authority was received by Lt. Colonel Teague to patrol these areas. The high ground was found to be unoccupied, and by 1600 hours it was occupied by elements of the Third Battalion. From here leading elements of the Third Battalion advanced meticulously through the western edge of Prum. By midnight, one-quarter of the town was secure.

While the Second Battalion was in Prum, it maintained its CP in the rear of a four story former department store which was subjected to constant enemy shelling from higher ground.

One shell had come particularly close to the front of the building when someone was heard running toward the CP at full speed. It was Captain Stephen J. Sanders. He skidded to a stop before Major James Burnside and said, "Hey, Major, did you hear that last shell?"

"Sure did," replied Burnside.

"Well, sir," said Sanders, "you can now see all the way through to the ladies-ready-to-wear!"

At dawn on the 12th, the Third Battalion continued to clean out Prum. This was accomplished by dark and one light counterattack had been repulsed in so doing.

Summary: In the nine days of offensive action, 4th through 12th of February, CT 22 had advanced approximately 10,000 yards through the Schnee Eifel; this advance included the breaching of the Siegfried Line, taking of Brandscheid, and moving over extremely rugged terrain to seize the Eifel focal point of Prum. Adverse weather conditions prevailed throughout the period; there was either snow or rain and sometimes both; and roads, either icy or muddy, were invariably mined. Initial enemy resistance was from the 326th and 340th Volksgrenadier Division, and 1082nd Security Battalion. However, after the Siegfried Line had been breached, the 2nd Panzer Division was rushed to this sector on February 7th and 8th. In the hard fight for Niedermehlen on the 9th of February, our Second and Third Battalions destroyed the First Battalion, 304th Panzer Grenadier Regiment of the Second Panzer Division. The CT took 12,373 prisoners during these operations; and approximately 150 Siegfried pillboxes and bunkers were reduced.

* * * * *

From the **13th to the 26th of February, 1945**, the Combat Team maintained defensive positions west of the Prum River. For the most part, combat activity was limited to consolidation of positions which overlooked the town of Prum and provided direct observation on the streets of the city. Consequently, movement in the town was restricted, and re-supply had to be carried on during the hours of darkness.

The CT defended the sector with two battalions along a main line of resistance, which included the ridge west of Prum and the dominating terrain north of the town. Defensive positions were shifted and improved throughout the period. Patrols, combat and reconnaissance, were constantly used along the Prum River. The battalions were rotated on line so that each had time to re-organize, re-equip, and train while assembled in reserve at Gondenbrett and Obermehlen. Effective fighting strength had been seriously reduced in the Siegfried operation, but gradually the combat efficiency was renewed.

An order to attack across the Prum River came to the Combat Team on the 27th of February. Preparation for the attack continued that night, and at 0315 hours the 28th of February, the CT attacked to force a bridgehead across the Prum River and to seize the town of Dausfeld. Dausfeld was situated in-between two hills which had to be seized in order to gain control of the town. One hill was approximately 800 yards northeast, the other hill almost 1,000 yards to the southwest: the former to be Objective 3, the latter Objective 4. The 1st and 3rd Battalions attacked abreast, the 1st Battalion on the south at Prum.

The 3rd Battalion forced a crossing of the Prum River about 2,800 yards northeast of Prum and seized Objective 3. Part of Company 'I' entered Dausfeld and four men were captured, so the remainder of the company did not attempt to enter until armored support arrived. Before noon, some of the armor had managed to cross the river, and within three hours this armor, supporting the 3rd Battalion, drove into Dausfeld. Company 'L' had captured several prisoners, and Lt. Arthur, after brief questioning, determined that they were a part of Hitler's youth movement and possessed a cocky attitude. When later questioned by the I. P. W. teams, they continuously spoke of the new secret weapon which was going to later defeat the Allies. All of them were paratroopers, but a number had received no paratroop training. A few of these prisoners had even been brought down from Norway as replacements for the Westwall.

The 1st Battalion, meanwhile, in a column of companies, crossed the river on improvised foot-bridges approximately 1,600 yards northeast of Prum. The 1st Battalion, suffering numerous casualties from mine fields, drove on and by noon had seized Objective 4, repulsed an enemy counterattack, and dug in.

Maj. Henley directed the 2nd Battalion across immediately behind the 1st and turned south to clear the ridge line opposite Prum. This action was only partially completed by dark, and the 2nd Battalion was forced to dig in for the night. Enemy forces here were determined elements of the 5th German Paratroop Division and resisted fanatically.

Supply and evacuation of the wounded handicapped the CT as enemy fire delayed the bridging of the Prum River for vehicular traffic until after dark on the 1st of March. In spite of this hindrance, the attack was resumed the next morning by the 2nd Battalion. This battalion was to clean out the wooded slopes east of Prum. These hills were found to contain trenches five feet deep with built-up firing positions. Observation from these trenches was so clear that it was possible to see men moving about inside of the houses in Prum.

This was obviously the reason the units in Prum suffered so heavily in casualties from sniper fire. Communication wires extended throughout the trenches and to listening posts on the Prum River. The Germans hadn't been caught by surprise; they had been well prepared. By mid-day, the slope was clear, and the

2nd Battalion was relieved by elements of the CT 12. The 1st and 3rd Battalions remained in position and did not attempt to advance.

Throughout this period, quotas of men were daily pulled back off of the front lines and given a three day pass to Paris. When the deserving men were selected to go back to Paris, they inwardly felt as if heaven had temporarily spared them the horrors of the battlefield and allowed them a bit of rest. These men loaded on two and one-half ton trucks and in a matter of hours were out of the combat zone. In Paris there were electric lights, music, and girls. It was like a slice of heaven brought to earth.

For three days and three nights, men were allowed to come and go as they desired. Paris was gay, the people were friendly, and everyone had a grand time. The night clubs, burlesques, movies, and sidewalk cafes catered to the combat soldier. After the three day period was over, the men returned again to the front. There was little complaining; no tears were shed; these battle-hardened veterans knew that their job was as yet unfinished and the sooner that they completed it, the sooner they would be able to return to the freedom of the United States.

On **March 2, 1945**, Colonel Charles T. Lanham was assigned to the One Hundred Fourth Division as the Assistant Commanding General. Thus ended a colorful association between a great fighting commander and a great fighting regiment. The relationship had sometimes been bitter; there was no easy way to lose lives. But there had come to be mutual respect.

Buck Lanham had led the regiment in every successful battle engagement of the war except the Normandy assault. He was driving, demanding, and aggressive. He was an able tactician, possessed a thorough knowledge of men, and commanded the respect of men who served under his command.

It was a brilliant and successful relationship and one the men of the regiment were destined to remember.

It was with genuine pleasure the regiment learned that Buck Lanham was to be succeeded by the regimental executive officer, Lt. Colonel John F. Ruggles, rather than by an officer from outside the regiment.

The 2nd day of March, the attack moved ahead, the First and Third Battalions fighting abreast. This terrain was fast becoming suitable for tank operations—the hills extended in long rolls, woods were scarce, and the bright sunshine was slowly drying out the mud. Before long the armored divisions would be able to smash ahead and rapidly crush the remaining German armies.

The 3rd Battalion advance was slowed initially by heavy enemy rocket fire and later by necessity of covering its left flank, which was now exposed due to the failure of CT 8 to take Weinsheim. The 44th Field Artillery Battalion laid down a perfect smoke screen between the town and the 3rd Battalion. The road leading southeast from Weinsheim was severed at 0808 hours, However, this drive was halted after penetrating approximately 1,000 yards east of the road; fire from both flanks covered all routes for further advance.

The 1st Battalion also received heavy enemy fire, and its attack east-ward was temporarily delayed. But, by 0630 hours the attack was underway; the advance ceased when four supporting tanks were destroyed by direct fire from the enemy. Thereafter there was bitter indecisive fighting at Bruhlborn, but by dark the situation was favorable.

As leading elements were receiving effective fire from both flanks as well as the front, CT 22 adjusted its position and secured for the night by diffusing the left flank in the vicinity of Weinsheim and slightly withdrawing the most forward element to better defensive terrain. Orders received during late evening placed Weinsheim as an objective of the Regiment,

In an attempt to gain surprise, the 3rd Battalion, shortly before dawn on March 3rd, moved to positions from which it might assault Weinsheim, Before the attack got underway, white flags were seen in the town, and patrols verified the fact that the enemy had withdrawn from there. Immediately the CT pushed out and seized the high ground around the town. This was rapidly done as the 11th U. S. Armored Division had been waiting for a suitable jump-off line; this was it. The 11th Armored moved through CT 22 and attacked east. At once, CT 22 received a new zone of action; the direction of attack was now northeast.

A coordinated attack by CT 22 and the 11th Armored Division on the 4th of March was delayed awaiting the seizure of Gondelsheim. At 1015 hours, the attack got underway, and by dark the Team had advanced 6, 000 yards to the high ground just east of Schwirzheim.

The following day, the 5th of March, Duppach was secured against only slight resistance. The 2nd Battalion, pushing out from Duppach, encountered very stiff resistance on a hill nearly 1,000 yards to the southeast thereof. By dark the hill had fallen. This hill formed the western edge of a saddle which had to be opened for the 11th Armored to pass through. The hill forming the eastern edge of the saddle fell to CT 12 shortly before dark that same day.

In agreement with orders, the 3rd Battalion relieved elements of the 11th Armored Division in Ober Bettingen and in the one-half mile bridgehead east of the Kyll River during the night of 6-7 March. At 0215 hours the relief was completed. The 2nd Battalion had moved into the right portion of the bridgehead, the 3rd Battalion in the left. Patrols of both the forward battalions from these positions moved into Hillesheim and encountered no organized resistance.

As the troops moved up the long hill from the Kyll River into Hillesheim, they anticipated sudden destruction. It was entirely too quiet. All along the way through the woods and along the roads could be seen the tremendous tank traps and connecting trenches. These trenches lined the hillside in a zig-zag pattern. Obviously their construction was very recent as the soil was newly turned. Large bunkers were found lined with timbers twelve to eighteen inches in diameter and covered with a thickness of several feet. Evidently these had been constructed by the German slave labor groups who were continuously working behind the front lines preparing the defenses for the withdrawing troops. The road into Hillesheim had been destroyed and blocked by a fallen railroad trestle. All along the railroad embankment it was possible to see foxholes which had been dug in preparation for a defense of the town. These were never used. The enemy was withdrawing too rapidly to get organized.

By midnight, both battalions had moved into and secured Hillesheim to include the high ground to the east. There had been virtually no resistance, much to the amazement of all concerned, and the capture of the town offered a covered shelter for the night. It was almost too good to be true.

Until the 12th of March, the Regiment held and consolidated its positions in the vicinity of Hillesheim. Patrols were sent out to the east and northeast to ensure security. Task Force Rhine, composed of Combat Team 8 and the 70th Tank Battalion, passed through the 22nd Regiment at 1100 hours on the 8th and moved rapidly eastward. In the ensuing days, the rifle battalions, assembled in Hillesheim, took advantage of the situation to repair and clean equipment, re-organize, and relax.

Late in the day of March 12th, in accordance with orders, preparations were made to move the Combat Team by rail and motor to the general area of Luneville, France. The move was made on the 13th and 14th of March. The two trains carrying the foot elements arrived in Bayon, France, in late afternoon of the 14th. From this point non-organic transportation carried the troops to their billets near Magnieres. The motor column moved through Prum, Bleialf, Dasburg, Luxembourg City, Metz, Toul, Bayon, to the respective billet areas. All elements of the Regiment closed in to the new location by midnight. Upon arrival, CT 22, as part of the 4th Infantry Division, became a reserve element of the Seventh U.S. Army.

After the move into France, the Combat Team, as a part of the 4th Division, had been engaged in continuous contact with enemy forces for 199 consecutive days. This well-deserved relief was fully enjoyed by each man. The climate in southern France was similar to that of Florida, and the men relaxed in the warm sun to take sun baths and swim in the nearby streams. It was unspeakably wonderful!

22nd Infantry heavy machine gun team near Prüm, Germany February 1945 — Photo from the 22nd Infantry Regiment yearbook published in 1946

22nd Infantry entering Prüm, Germany February 1945 — U.S. Army Signal Corps photo

16. The Enemy Resistance Crumbles — Bavaria

"It is with the highest praise that I commend the officers and men who so gallantly fought the battles at Lauda, Konigshofen, and Bad Bergentheim. Words of praise could never express to each individual the sincere appreciation I hold within my heart for this great victory."

— *Lt. Colonel John Ruggles*
15 March 1945.

Upon termination of the period in which Combat Team 22 was an element of the Seventh U. S. Army reserve, and during which time the CT had again raised its combat efficiency by training, motor maintenance, and rehabilitation work, the unit was again committed to action against the Germans on the 30th of March 1945. The Twenty-Second Regiment, Fourth Division, was not to be an important element in the final thrust to defeat the Germans. After crossing the Rhine, the main objective was to attack to the Brenner Pass. It was established that the Germans were planning to stage a delaying action until their forces from northern Germany could form a defensive line in the Bavarian Alps in conjunction with the troops which they had fighting in the Po Valley and northern Italy.

The Brenner Pass was the only logical route through which

the forces might again contact. This pass must be seized as rapidly as possible in an effort to prevent a uniting of forces and to isolate these enemy units in Italy from those in the Bavarian Alps and the rest of Germany. Though the regiment never actually reached the Brenner Pass, it did in approximately thirty-seven days disrupt enemy communications, transportation, supply and in turn force them into an utter state of confusion. The regiment in the month's drive attacked and drove the Germans back to the very edge of the Bavarian Alps in southern Germany. In order for this mission to be accomplished, objectives were sent down daily to the CT for its continuous drive to the southeast.

On the 30th of March, the regiment staged forward in rear of the 7th Army advance. Movement began by motor convoy about 0030 hours. The regiment was to move to an assembly area east of the Rhine. The route to be taken was via Neustadt, Baddurkheim, Heuchetheim, Worms, Heppenheim, Laudenbach. The column closed in shortly after dawn after travelling fifty miles, but immediately upon arrival all units were informed that the stop was to be brief and a further move would be carried out that day.

Motorized patrols moved out at once and were followed by the main body. Because of poor roads and traffic congestion, the advance was slow, but forward elements pushed on with the mission of passing through the 12th Armored Division and securing the Army bridgehead thirty miles east of the Rhine. At dark, the Regiment secured for the night in the environs of Schonmattenwag, Hirschhorn, and Finkenbach.

The advance east was resumed early the morning of the 31st, with the mission of securing Eberbach. Preceded by strong motorized patrols, the attacking battalions easily reached these objectives with only slight resistance. This advance covered six miles.

Elements of the 12th Armored Division and the 101st Cavalry Group, by their operations to the east, were greatly facilitating the advance of the Regiment.

* * * * *

The Combat Team mission on the 1st of April was to secure a crossing of the Tauber River, about 35 miles to the east, and secure an objective near Grunsfeld. The Regimental I & R Platoon reconnoitered routes for about twenty miles east and encountered units of the 12th Armored and returned. At 0600 hours, the Regiment, motorized only with organic transportation and several medium tanks, attacked east. The 2nd Battalion was completely motorized in order to affect a rapid advance. The unit pushed rapidly through Mudau, Buchen, Hettingen, Gerichstetten, Buch Ahorn, Heckfeld, and Lauda. The latter part of this move was made on foot against light resistance. Against in-creasing artillery and mortar opposition, the unit crossed the Tauber River during the early afternoon. By late afternoon the 2nd Battalion had secured Grunsfeld and the high ground around the town.

In the meantime the 1st and 3rd Battalions had been shuttled forward, and both battalions attacked through the 2nd Battalion. The 1st Battalion pushed slightly more than 1, 000 yards to the southeast of Grunsfeld and seized Krensheim.

The 3rd Battalion, moving through Lauda after dark, was struck by German fighter planes. Because the head of the column was not affected and the rear was, the column split in Lauda, and contact was lost. Capt. Reid, of Co. 'I', took command of the last section of the column and deployed the vehicles off the road to avoid strafing or bombing. In the eastern edge of Lauda was a railroad overpass, and men huddled silently beneath it as the

German planes circled overhead looking for a target. In the distance huge flames lit up the sky as the small town of Konigshofen burned. The planes could be heard firing their machine guns and blue streaks lined the sky as the incendiary bullets flashed forth in their mission of destruction. As always, some irresponsible GI lit a cigarette and the match flame flickered in the wind. Immediately shouts went forth, "Put out that damned cigarette. " It was too late; the planes thundered down, strafing the road and lighting up the area with blue light.

After a short while, the rumbling motors were gone, and the column quickly moved down the road again. Each man knew the planes had only gone for a re-supply of ammunition and bombs, to return laden with their destructive missiles. Farther down the road, the column passed the area occupied by the 44th Field Artillery Battalion, and the planes were again heard. At once the 377th Antiaircraft Bn. protecting the field artillery opened fire, and quick flashes of ack-ack burst in the sky. The column continued to move in spite of the aircraft, and several hours later arrived in the prearranged assembly area. During the day the Regiment had moved over fifty miles through very hilly terrain and over poor secondary roads.

In comparison to the previous encounters with the enemy forces, the past few weeks had been easy, casualties light, and the advance quick and continuous. This brought with it a new hope that Germany would soon collapse and the European War which had begun years before would finally cease.

There was ground yet to be gained, battles to be fought, and victories to be won. On the 2nd of April the 1st and 2nd Battalion were attached to Combat Command 'R', of the 12th Armored Division. These two rifle battalions attacked to the south to clear the east bank of the Tauber River and held the armored command across the stream. Enemy artillery, rocket, and small arms fire was moderately heavy, but Marbach, Kutzbrunn, and Hofstetten were captured. Opposition to this attack to the south indicated a sizeable enemy force north of Bad Mergentheim and east of the Tauber River, CT 22, with only one rifle battalion remaining under its control, searched through, and mopped up its objective with motorized patrols.

Before noon the 3rd of April, the 1st and 2nd Battalions returned to the control of CT 22. The 3rd Battalion reverted to reserve and moved to an assembly area in Gerlachsheim. The day was utilized in position consolidation and the re-grouping of forces.

Well observed, heavy enemy rocket, mortar, and artillery fire on for-ward elements of the 1st and 2nd Battalions further indicated the general enemy defensive build-up south of the Combat Team.

Shortly after daylight, an enemy group of more than forty men slipped in between Companies 'B' and 'C', and a fight ensued. Three jeeps attempting to carry supplies to Co. 'B' were ambushed by these Germans and two men were killed, three wounded. Around 1430 hours, Co. 'F' with a detachment of tanks was sent down along the eastern edge of the woods to contact 'C' Company. Capt. Surratt, Battalion S3, was killed by a mortar shell just after he had succeeded in once again tying in the front lines of 'B' and 'C' Companies. Fighting was bitter, but within an hour, 'F' Company had managed to extricate Company 'C'.

An attack to the south on the following day was delayed pending the completion of plans and the issuance of orders. Shortly after noon, the 3rd Battalion attacked through the 1st Battalion across the Marbach-Hofstetten Road. Initially, progress was rapid, with only slight opposition, and Co. 'L' reached the edge of the woods about 1,100 yards northeast of Konigshofen, where it held, as Companies 'I' and 'K' pushed west to clear the wooded hill 300 yards right of Co. 'L'. Heavy small arms resistance was met on this hill, and the advance stopped. The command group from Co. 'M' and the 81mm mortar platoon mov-ed into the town of Konigshofen and secured it for the night.

In talking to a couple of civilians living in the town, it was learned why the town had been burned. German SS Troops several days before had found civilians possessing the white flags of surrender, which they intended to display upon the arrival of the Yanks. The SS Troops immediately poured gasoline on the houses and set fire to them, telling the civilians that they would be shot if they ever attempted to surrender. Hundreds of people were homeless, and children gathered in the school building, crying that their homes and parents were gone. These were the houses seen burning from Lauda several nights before.

The 1st Battalion had, in the meantime, adjusted its positions to the rear of the 3rd Battalion, and the 2nd Battalion probed to the southeast in order to attain a position to the left rear, northwest, of the 1st Battalion. Before dark, the 2nd Battalion had secured the town of Messelhausen.

The 3rd Battalion, which was meeting the stiffest resistance on the Konigshofen Ridge, was held to very small gains on the 4th and 5th of April. Heavy woods prevented the use of tanks in the attack, and intense, accurate small arms fire met every movement of the foot troops. As Capt. Reid later stated, "That was the doggonedest small arms fire 'Big Item Co.' will ever encounter". The ground on this ridge consisted of the heavy woods with rock piles parallel to the line of advance with holes cut in them for German automatic weapons. Almost every casualty suffered was by a direct hit in the head with one bullet. Lt. Neel, who led his machine gun platoon all the way from Northern France through Germany, was quite correct when he said, "In a battalion there are two machine gun platoons, one of which is always attached to each assault company, and there are always two assault companies, so while one rifle company is pulled into reserve to rest, the machine gun platoon is attached to the next assault company. In other words, a damn machine gunner never gets a rest."

The enemy were found to be paratroopers and as fanatical as any troop of German soldiers we had fought. They were young, from fifteen to nineteen, but they fought with a fanaticism of which we had read, but seldom had met. It was this group of men that against Co. 'C', had led one of the old-fashioned charges in which an officer stood up, and shouting and yelling, the men had charged forward, only to be cut down and hurled back by our machine gun, and finally by hand-to-hand bayonet fighting.

One of these young Hitler youth had a wound in his leg which prevented his crawling away, and he was sitting under a tree. In the clean-up of wounded after the attack had been repulsed, the battalion medics, as usual, were going from person to person, tagging them, administering first aid, evacuating litter cases, helping the wounded, and sending them to the safety of the rear areas where they could be treated by the surgeons. One of the medics started to approach the wounded German lad when the boy picked up a "potato masher", the common name for the German hand grenade, so called because it was shaped exactly like the old-fashioned wooden potato masher, with long wooden handle, for throwing. The medic stopped and pointed to the red cross arm band which he was wearing, then to his medical kit, then to the German. The German stared at him stonily, and as the medic again moved to approach him, he unscrewed the cap of his grenade and the medic hit the dirt, expecting the German to throw it. Instead, the lad held the grenade immediately under his chin until it went off, blowing his head completely and cleanly from his body. Such was the fanaticism of the Hitler youth.

Contact between our 2nd Battalion, still in Messelhausen, and CT 12 was established on the 6th of April, and, in agreement with plans, the 2nd Battalion reverted to regimental reserve at the time. As CT 12 pushed south, the 1st Battalion attacked and cleared the woods 1,000 yards north of Deubach and later advanced into the town itself.

The 3rd Battalion continued to meet heavy small arms resistance during the morning hours. About mid-day, the fire slackened, and a strong combat patrol was sent forward; this patrol returned with the information that the enemy was apparently withdrawing to previously prepared defenses around Bad Mergentheim. Patrols from the 1st Battalion had worked into the woods just south of Sailtheim and had heard voices and digging in various parts of the woods. Elements of the battalion were sent through the woods with tanks "firing like hell". 'C' Company attacked on a cross-country maneuver and captured Deubach while Co. 'B' advanced into the town shortly thereafter by coming down from the north.

The 2nd Battalion moved forward at approximately the same time as the 1st and seized Oberhalbach and Loffelstelzen.

The 3rd Battalion purposely delayed and then started cross country to Unterbalbach. As they left the Konigshofen Ridge, the German dead lay piled like cord wood over every conceivable defensive terrain feature. Some of the Germans had been dead for several days, and their skin was turning black, and the blood clotted clothing swarmed with flies. There was the foul stench of death in the atmosphere, while overhead a single bird chirped as if in mockery to the mortals below. Proceeding on across the open fields, the Battalion moved into Unterbalbach and Edelfingen. By night the CT was overlooking Bad Mergentheim from the north.

At once the 2nd Battalion sent a patrol into Bad Mergentheim to investigate and determine the enemy strength. Lt. Kornreich, of 'E' Company, led the patrol in about midnight. He was accompanied by Capt. Herrick, 2nd Bn. medical officer, who was to secure the release of three wounded American soldiers in the hospitals there. The patrol returned with the three wounded Americans and information that there was no organized resistance in Mergentheim and that the vehicular bridge there still remained intact. With the information, the 2nd Battalion sent a larger patrol into Bad Mergentheim to penetrate through the town and reach the high ground to the south. This patrol returned with the information that the enemy was digging in south of Hill 307. Too, much to the surprise of the men in the patrol, the Burgomeister (town mayor) stated that the German civilians who lived in the town had, after a lengthy discussion, persuaded the German soldiers to surrender or to withdraw. The civilians were also responsible for keeping the bridge intact. These people realized that if the soldiers attempted to defend the town, American artillery, mortar, and tank fire would completely destroy both themselves and the town.

Lt. Col. "Lum" Edwards, Regimental S3, first received this information and immediately reported it to Lt. Col. John F. Ruggles, Combat Team Commander, who then ordered the Regiment to move into Bad Mergentheim and outpost the surrounding terrain. This action was initially carried out by the 3rd Battalion. Erroneously, a 3rd Battalion wire team jeep carrying Lt. Rose and his two forward linemen struck out down the main road into the town. They crossed the bridge and drove down across the railroad tracks into the center of town. Upon arrival at the town square, no GI's could be seen, and Lt. Rose turned to the driver and whispered, "My God, man, we're the only ones in town. Turn this jeep around and let's get the hell out of here; we might get killed."

The wide-eyed occupants of the jeep quickly turned around and sped down the narrow streets. German civilians stared in amazement at the four passengers, not realizing their mistake. As the jeep left the edge of town, they met the lead scouts of one of the rifle companies moving toward the town. As the jeep drove by the advancing foot troops, the driver ironically leaned out of his jeep and said, "There's no one in there; we just cleaned out the town."

The 3rd Battalion moved on into the town, leaving 'I' Company to guard the hospitals, and the balance of the battalion moved on to Hill 374 southeast of the town. Bad Mergentheim was a hospital center and recuperation area for over a thousand Germans. Many homes had been converted into hospitals to care for the wounded.

The 2nd Battalion, following the 3rd Battalion, moved through the town to Hill 307, turned southeast, and moved to Hill 374 abreast and east of the 3rd Battalion.

The 1st Battalion had previously assembled in Oberbalbach and just before dark moved to Gerlachsheim as division reserve.

In conversations with the civilian populace of the town, it was not a rare occurrence to speak entirely in English. The German youth had been taught English in their schools as long as they could remember. Hitler had been preparing for this war and the defeat of the United States many years. The town of Bad Mergentheim was a resort town somewhat similar to Hot Springs, Arkansas. It was situated along the Tauber River in between the surrounding hills. The streets within the town were narrow and the buildings and homes closely fitted. The buildings were not of large structure, usually only two floors with windows and large verandas opening into the streets. There were hotels, government buildings, and hospitals throughout. In the edge of town and on the hillsides, large mansions could be seen obviously owned by high government officials or business executives who came to Bad Mergentheim to relax and better their health. After the arrival of the Americans, life continued normally with only minor restrictions and curfews in effect. The German wounded remained within the hospitals, but German medical officers could be seen moving through the town on their routine check accompanied by a couple of our medics. In general, the people were very cooperative and quite thankful that their town had been spared the ravages of war.

22nd Infantry near Scheuren, Germany March 6,
1945 — U.S. Army Signal Corps photo 324563

17. A City Dies And Remembers Jones

After enjoying a few days of luxury in Bad Mergentheim, the Combat Team forged ahead. The German armies were in such an utter state of confusion that they could only fight delaying actions, but every wood, hilt and town was a virtual strongpoint containing a small group of enemies who would fire enough to temporarily delay the unit.

Every day the huge silver army transport planes could be seen soaring overhead loaded with supplies for the armored columns spearheading the drive. It was an impossible task for the trucking and quartermaster companies to keep up with only bare necessities, because as long as the enemy was backing up, they knew the battles to be won were becoming fewer.

The landscape through Southern Germany was slow rolling hills spotted with small patches of woods and thousands of tiny villages. The towns were usually no more than a mile apart with roads connecting from all directions. The atmosphere sparkled with spring, and life returned to the trees. Over-head birds could be seen, and white billowy clouds lazily floated onward. Then suddenly the beautiful countryside would become a mass of bursting flame, black smoke, and shrieking metal; machine guns would drown out the sound of the birds; in the distance the deathly cough of mortars could be heard; and the powerful motors of the American tanks grumbled as they moved into firing positions. The enemy wanted to fight—so shall it be. In a few short hours the Combat Team moved on, leaving behind a hideous portrait of death and destruction.

On April 20, 1945, the Combat Team, after a move of approximately 55 miles through the towns of Weikersheim, Ober Rimbach, Weiler, Rote See, and Beuerlbach, arrived upon the high ground overlooking Crailsheim from the north. Enroute numerous skirmishes were fought, but the advance was so rapid it would be repetitious to describe each encounter.

The First Battalion, under the command of Lt. Colonel George Goforth, led the advance on the 20th by moving into the northeast corner of Crailsheim. Here serious opposition was encountered from light artillery and sniper fire. One of the things which Americans will never forget was the order requiring every German house and building to fly a white flag or be destroyed. An advance patrol was usually sent into the town ahead of the combat troops and artillery, and the Burgomeister was informed that if the Germans defended the town or if American soldiers were harmed within its precincts, it would be burned to the ground. If, on the other hand, the civilians did not interfere and force the withdrawal of German troops to the other side of the town, it would be spared. In almost all cases, the Germans of Bavaria were anxious to have their homes spared, and the white flags went up at once. Usually a time limit of thirty minutes was set. It seemed strange, in the light of hysterical and arrogant speeches made by the Nazis through the years, to observe the servility of most of the civilians, although perhaps their reaction was only human.

It was now apparent that the once mighty German Army was disintegrating rapidly, and everyone, including the Germans themselves, realized they were completely defeated; still they fought on, a fact which we mightily resented, because it made every casualty we took and every casualty they sustained so completely useless. It increased our bitterness.

The First Battalion was ordered to take the town of Crailsheim, a city of perhaps some eight or ten thousand inhabitants. The Battalion drew up its ring of steel in preparation for the attack on the northern edge of the city and there stood fast while Lt. Jones was ordered with two enlisted men to approach the city officials and arrange for its surrender, by Lt. Colonel George Goforth, the Battalion Commander. This Lt. Jones immediately did, and upon approaching Crailsheim with an interpreter, he contacted the Burgomeister and gave to him the American terms for the surrender of the city. The terms had apparently been received and accepted when, without warning, the three American soldiers were shot in the back and killed.

When this information was given to Colonel Goforth, he issued orders with tenseness and a concealed fury which communicated itself to every man in the battalion. Instead of ordering that the city be attacked, he called for artillery fire, and he personally adjusted the mortars, ordering them to fire white phosphorous and to burn every building in the city. These orders were carried out, with Colonel Goforth observing. If occasionally the smoke and flames blew away and a building was revealed still standing, the Colonel adjusted the mortars and soon the building was in flames. The city of Crailsheim will long have cause to remember a Lieutenant named Jones and his funeral pyre. It was a cowardly and needless gesture on the part of the Germans; it was a furious and just retaliation from a combat battalion which had learned to hate the treachery of a defeated enemy. A city died, and remembered a Lieutenant named Jones.

As the city poured forth the black smoke of destruction, the First Battalion moved into the town. They now wanted to fight, to find the treacherous cowards who had only minutes before killed three men of the battalion. Company C, with three tanks, led the assault; Company B followed. This time there were no lead scouts as each man wanted to be first. Street fighting had long ago become a perfected knack with these men. They quickly paired up and moved ahead, meticulously searching each possible hiding place for snipers. Darkness fell, and the men, still boiling with anger, hunted the enemy by the light of the burning buildings. The night was overcast, and the white clouds reflected the firelight, illuminating the narrow streets. The population of the city knew that they had made a mistake and were paying for it.

While the fighting was going on in Crailsheim, the Second and Third Battalions moved up to secure the flanks and rear. The men of these battalions could clearly see the burning reflection and they, too, understood what had happened.

The Combat Team didn't remain long in Crailsheim; their job was ahead. "Task Force Rodwell," commanded by the assistant division commander, Brigadier General James S. Rodwell, passed through the CT the next day and temporarily led the advance.

This Task Force greatly facilitated in the further advance of the Combat Team for several days as it drove the Germans back or surrounded them, forcing a surrender.

22nd Infantry enters Crailsheim, Germany, April 21,
1945 — U.S. Army Signal Corps photo 204179

18. Love Thine Enemy—An Interlude

Otto Oehring was a quiet technician fifth grade who served as Chaplain's Assistant to Chaplain Boice. He was considerably older than the average GI, distinctly German, and speaking English with a strong German accent. It was also easily apparent that he had succumbed to his ancestral tradition of baldheadedness.

Every attempt had been made to get Otto transferred to a station complement or a service command prior to leaving the states, but to no avail. The chaplain had learned that Mr. Steinway of New York, in one of his annual tours through Germany, had passed through the town of Heilbrounn, home of the Glass Piano Company, where he had picked Otto as a master piano craftsman, building of pianos, and had taken him to New York to help Steinway build their fine pianos.

And so Otto, two sisters and a brother had left Germany and had settled in New York, becoming good and reasonably prosperous American citizens. The remainder of his family, we soon learned, consisted of an aged mother, a crippled sister, and three brothers, two of whom were officers in the German Army.

In the final campaign for Southern Germany following the swift move across the Rhine River of the Division, the Regiment was ordered to seize the town of Bad Mergentheim, It was at this point we noticed Otto growing restive. This was so completely unlike him that we tried to fathom the reason, and soon discovered it was because we were a mere seventy kilometers from his home. He told us there had been no word from his mother, sisters, and brothers since the day Pearl Harbor was bombed.

Late one evening the chaplain gave Otto instructions to get his work out of the way and to be ready to make a quick trip the following morning. The next morning Mr. Mitchell, Army Red Cross Field Representative, the Chaplain, Otto, and Jonesy, the faithful driver, started toward the town of Heilbrounn which we knew had been first bombed, then fiercely fought through, and finally "liberated" by American troops. The trip was uneventful but long, because all bridges had been knocked out and it was necessary to travel great distances in order to find fording places or bridges still intact. In the middle of the afternoon the party arrived on the outskirts of Heilbrounn. Otto tried to keep from showing his nervousness, his anticipation, and his fear without success. He knew as did we that Heilbrounn had been almost bombed out of existence by American heavy bombers on the 28th of December, 1944.

We came into the city from the east, headed over the ridge down into the valley in which the city was located. We held our breath and then suddenly the shockingly appalling sight stunned us. Heilbrounn as a city had ceased to exist. Blackened, devastated piles of ruins stood before us. American bulldozers had pushed the debris from the streets, filled the shell craters, and what the bomb-ers had not ruined, artillery completely finished. It was incredible that any-thing could have remained in the city during this period and lived.

We drove on down through town and there was not enough

of the city left for Otto to get his bearings. We cruised around for some thirty minutes, finally coming by chance close to the old section where the Glass Piano Factory had been, and then, we found his home. We stopped the jeep and went quietly behind Otto as he went past the little garden and the statue of Pan piping by a little fish pond, to the door. The glass of the door was broken out, but in all of Heilbrounn there were only two blocks of buildings left standing and this house was in the two blocks. There was a little metal sign on the door which said, "Mrs. E. Oehring", but the house seemed cold and was obviously lifeless. The door was unlocked, and we went through. There was nothing there—no furniture, no living person, nothing, nor was there any trace of the family or where they could be found. Otto wanted to give up then, perhaps because he was afraid of the truth and what he was sure he must find. Having heard voices from one of the cellars next door, we persuaded him to talk to some of the civilians, and to ask them if they knew where his family had gone. He explained quickly that his aged mother and crippled sister had been in the basement of the house all through the bombing but that his brother had come back from Karlsruhe following the bombing and had taken the family to a little town called Kirchausen which was only ten kilometers away.

We wasted no time in moving on west through the city under the ruined and twisted steel of the railroad overpass and west toward Kirchausen. It seemed strange to see this lovely German village completely untouched by war after the ruins of Heilbrounn. We inquired first at the drugstore, but the druggist was obviously too much upset at the sight of the hated American soldiers in his store to be of any practical assistance, so we went direct to the Burgomeister, who is always in touch with every event which affects the life of his village. The Burgomeister, not only aware of his own responsibility but of the power of the American Army, readily informed us that he knew where the family was living and so took us to the other side of the village to the family of Mrs. E. Lambert.

On the east edge of town close to the green gently sloping fields there was a small stucco bungalow with bright red shutters and hollyhocks growing in the yard. The first sight that greeted our eyes was a comely young German woman with darling blue-eyed twins in either arm. She stared at us for a moment with a highly frightened look on her face as we came through the gate and passed the hollyhocks toward the tree under which she was resting. Then she saw Otto — in American uniform, true, but still with a resemblance to the Oehrings that was unmistakable.

She simply whispered softly, "It's Otto! I know it's Otto!" Otto kissed her and hugged the twins of whom he had not even known. She guided him gently toward the door and we stood back, not wishing to intrude on this moment for these people whom a war could not separate. Inside the door the house was small but spotlessly clean and comfortably livable. We turned through the hall to the sitting room at the right and there, sitting quite alone, was one of the most distinguished-looking women we had ever seen. Her face was lined with wrinkles. Her eyes were China blue, and her head was crowned with luxurious waves of snow white hair. She wore a cheap black dress as if it were royal purple, and her hands were quietly folded as she sat and rocked and dreamed of a better day. And then she looked up.

Her quick eyes registered no surprise. She stood and held out her arms and cried, "Otto, Otto! Gott mitt uns! Mine Otto!" And then — we never quite knew how, for our eyes were dimmed with tears... Otto was in her arms, her boy again, not an American soldier in an enemy land, just her boy. It was a moment for them alone, and we went back out into the sunlight and stood by the hollyhocks, wondering at the sordidness of war and at the universality of love. Surely this was in the best tradition of Him who said, "Love thine enemy."

Chaplain Bill Boice with his driver Otto Oehring

19. German Defeat — Final Action

"Full Victory in Europe has been attained."

— *Gen. Eisenhower*

The Combat Team, after leaving Crailsheim, continued to drive on to the southeast. Strong motorized combat patrols preceded the main body and located enemy pockets of resistance and reported that these could be destroyed.

All along the route of advance, German vehicles were found, some of which were still in good running condition. Capt. Crawford, of Antitank Company, and Lt. Milhous, from 'M' Company, located two German buses, and converted these into mobile kitchens. Actually, the mess sergeant and the company mechanics worked these two buses over and anchored the field ranges to the floor, after removing the seats. By so doing, these kitchens were able to prepare the meals while the motor column was moving. The mechanics continuously worked to repair the motors and keep them running. Spare parts proved to be a problem, and the German vehicles apparently had no set standard of equipment or replacements. T/4 Don Wensink, of 'M' Co., who was by nature easy going and seldom infuriated, lost his temper more than once in trying to repair a German bus or truck.

One cold morning after traveling some half-mile through snow knee-deep in order to locate a truck, Sgt. Wensink, in a tone of complete disgust, said, "These Krauts never made two pieces of equipment alike. If I ever see another German truck, I'm going to blow the damn thing to pieces before someone wants me to get it to run." In the last few weeks of the war, the entire CT was usually able to ride on some type of vehicle because there were so many captured vehicles available.

The advance crossed the Kocher River through Zang into Heidenheim. Heidenheim was a beautiful city situated in the low, rolling hills of southern Germany. The 3rd Battalion was the first element of the Combat Team to move into the city. This they did the evening of April 24th, and by noon the next day the entire 4th Division had moved up into the city. The advance pushed on, and the CT crossed the Danube River near Lauingen on the 25th. The progress continued almost entirely uncontested until the Regiment reached the Lech River. Here the bridge was destroyed, and the Regiment assembled near Graben, awaiting completion of a bridge by the Engineers.

The only means which the enemy had to delay the advancing troops was the destruction of bridges which delayed the motorized columns. After crossing the Lech River, the CT proceeded to the Isar River and the bridge site just east of Unter-Schaftlarn. Here again the demolished bridges delayed the crossing. It was here that the Regiment captured a number of German women in the army uniform. When questioned by the IPW team, these women stated that they were typists and clerks for headquarters units. Enroute here, the Regiment had passed just south of Munich, and the tall smoke stacks of the city's factories could be seen in the distance. The 3rd and 45th Infantry Divisions were moving down the roads adjacent to the 22nd Regiment as they assaulted the city of Munchen. With the aid of field glasses, the white flags of truce could be seen hanging from the larger buildings. The 1st Battalion, which had the mission of blocking all roads from Munich into the Combat Team zone of action, secured important enemy installations to include a supply dump of 10, 000 liters of aviation gasoline and 100, 000 liters of fuel oil.

The bridge across the Isar River was not completed until the 2nd of May. Shortly after it was completed, the 4th Reconnaissance Troop, with Co. 'K' mounted on their vehicles, led out in the renewed advance. The 2nd and 3rd Battalions followed up and rapidly swept through their zones of action, meeting only scattered token resistance. As before, the advance was further expedited by the use of captured enemy vehicles. With infantry mounted on vehicles of the 4th Reconnaissance Troop, tank destroyers of Company 'C', 610th Tank Destroyer (SP) Battalion, tanks of the 70th Tank Battalion, and by using organic and captured vehicles, the assault battalions were entirely motorized.

The resistance just north of Gmund was the heaviest reported and consisted of small arms, mortars, and artillery fire. By late afternoon, the Combat Team's advance to the southeast covered over twenty miles.

The Third Battalion, moving into Miesbach just before dark, met slight resistance. Captain Roche, Pfc. Kaiser, and his driver were well ahead of the advance when their jeep was fired upon by Germans on either side of the road. Panzerfausts, the German bazookas, hit around the jeep and Captain Roche told the driver to "get the hell out of there." Pfc. Kaiser had fallen from the jeep and was temporarily stunned. Captain Roche and the driver continued down the road and found a road block, forcing them to abandon the jeep. Securing the machine gun, they returned through the woods to Miesbach and reported the incident to Major Kemp.

Several days later, Kaiser returned to the Third Battalion and gave this account: When he noticed the jeep was gone and that he was alone, it was too late to escape. The Germans had surrounded him and took his weapon. Kaiser, who had lived in Germany until 1938 before coming to the United States, spoke excellent German; however, he did not tell the enemy this; he told them that he had learned it in school. They took him to the next town to their command post where the commanding officer interviewed him and asked about the advance of the Americans. Several days later, Pfc. Kaiser returned to American lines and encountered elements of the 101st Airborne Division. He had persuaded some seventy-five Germans to accompany him and surrender.

The Combat Team, stopped for the night, deployed with the Third Battalion in Meisbach; the Second Battalion in Gusteig, Duenbach, Festenbach, and the First Battalion in Holzkirchen.

While the First Battalion was located temporarily in Holz-kirchen, word came to the Battalion Commander, Lt. Colonel George Goforth, that a German woman in the town was asking the Americans to take over her supply of liquor. Her husband had been a liquor dealer and had samples of almost every kind of liquor stored in the basement of his house. Vermouth, gin, brandies of every sort, cognac, wine, and champagne were among the abundant samples that were stored.

Knowing that the Americans were on their way, the husband had rounded up the Russian slave laborers and had poured concrete into walls over the entire cache, sealing it completely and making it look as if it were a part of the cellar, in order to keep it from the Americans. Now, however, having been liberated, the Russians remembered the supply of liquor which they had worked so hard to preserve and now had but one purpose in mind, and that was to obtain it. The woman had discovered that, armed with axes and sledge hammers, they were on their way to her home and, knowing the Russians and the German treatment of them, she had a pretty good idea that neither the liquor nor her home nor she herself would be spared, especially after the Russians had put some of the liquor inside.

She was therefore asking the Americans to come at once and to take over the supply. Needless to say, she did not have to ask more than once, and the order went out for the S-4 to guard and, if need be, confiscate the liquor supply to prevent the Russians from taking it, since it was the obvious duty of the battalion to preserve order.

Since the battalion did not have enough men to waste to keep a constant guard on the cellar, it was decided that the most sensible thing to do was to liquidate the liquor. It was also felt that the most practical way to do this was to drink it. When the liquor was picked up, however, the German woman impudently asked the lieutenant in charge if the American government would pay her for it and presented a bill which represented the equivalent of about five thousand dollars. The lieutenant, having been taught that a good soldier is never caught short, replied chivalrously and with no hesitation that he would be glad to sign such a receipt. He thereupon wrote out a complete and legitimate receipt and gave it into the hand of the woman, who was perfectly happy and satisfied. The receipt read: "This is to certify that the American Army has this day confiscated the liquor supply of this concern to the sum of five thousand dollars ($5, 000), this sum to be deducted from the amount of money which the German Government owes the American Government from World War I."

As the personnel of the Combat Team dug in for the night of May 2-3, 1945, little did they realize that this day had been the last day of actual combat for them in the European Theater of Operations,

At dawn the 3rd of May, the 2nd and 3rd Battalions swept through their respective zones with combat patrols but encountered no enemy. Division orders for the relief, movement, and assembly of the Regiment in the vicinity of Holzkirchen were received in the early afternoon. The Regiment was relieved shortly thereafter by elements of the 101st Airborne Division. The relief was complete by midnight. Totals compiled by the S2 Section showed that the Regiment had in the past two days processed 3,270 prisoners.

Orders came down during the night for a movement by the Combat Team to an area near Nurnberg, Germany. Immediately plans were drawn up, and at 0930 hours the 4th of May, leading units of the CT crossed the IP. The motorized column proceeded via the autobahn, superhighway of Germany, through Munich, Ingolstadt, to the assembly areas of Schwabach.

Upon arrival in the new area, planning was initiated, and preliminary orders were issued for the division of the area into separate battalion occupational sectors and for a thorough reconnaissance of the new area.

From **5-9 May, 1945**, the 22nd Regiment was engaged in a non-tactical occupational and guarding mission in the Nurnberg and Ansbach area of Germany. The Regiment had no contact with the enemy from the 5th of May until 0001 hours the 9th, at which time the European War officially ended with the unconditional surrender of all German naval, land, and air forces.

On the 5th of May, the Regiment divided its zone into four subordinate sectors of responsibility. Each of the three rifle battalions and the 4th Engineers, now attached to the 22nd, moved into and assumed control of one of these areas. The disposition of the units was as follows: Regimental Headquarters was in Schwabach, 1st Battalion Headquarters was in Nurnberg, 2nd Battalion

Headquarters was in Roth, 3rd Battalion Headquarters was in Spalt, and the 4th Engineer Battalion was in Weidach. These positions were maintained until May 9th. A training schedule was set up, and the troops carried out all types of training. Motorized patrols daily swept through the Regimental zone of occupation.

* * * * *

When the announcement of cessation of hostilities was heard, Chaplain Boice sent down a letter to the men of the Regiment to be addressed to their families and mailed home:

"This evening Admiral Doenitz has announced to the German people the unconditional surrender of all German fighting forces.

"Had this surrender occurred the 1st of September on our wave of optimism when we hit the Siegfried Line, or immediately after the defeat of Von Rundstedt and the successful crossing of the Rhine, we would have been wild with joy. The news of Germany's surrender was received by all of us with a calmness very nearly approaching indifference about the feeling deep within our hearts.

"There was no revelry last night, no drunkenness, no shouting, no flag-waving, no horns blowing; there was a sober realization that it was all over, at least so far as Europe was concerned, and that we, by the strength of our arms and by our own courage, had, with the help of God, completely and finally defeated everything that the warped and twisted soul of a perverted nation could hurl at us.

"We take no undue pride in what we have done, for we are sobered by the blood of our comrades which cries up to us from every foot of ground from Normandy to Berlin and from Holland to Italy. We have done what we had to do for you, as well as for our own peace of mind.

"I am proud of my officers and fellow soldiers in the 22nd Infantry Regiment. There is not one single fighting day of which we must be ashamed or for which we must make excuses. No regiment in the ETO has more right to hold its head high and to march with shoulders back, colors streaming, than this one. Its record, its casualties, its achievements, and the respect it instilled and the terror it struck in the heart of the German Army speak for themselves."

Lieutenant Colonel George M. Goforth, Commanding Officer, 1st Battalion 22nd Infantry; Lieutenant Colonel Earl W. Edwards, S-3 Operations Officer, 22nd Infantry; Lieutenant Colonel John F. Ruggles, Commanding Officer, 22nd Infantry — U.S. Army Signal Corps photo

20. The Men Who Were Not There

And so the war was over. It was a fact far too deep for us to grasp fully and we realized somehow that we should be more grateful than we were, that probably we should do all of the things which we were expected to do, like blowing horns and tooting whistles and perhaps getting drunk. But we didn't. We simply thought of the hundreds and hundreds of our friends who had given everything they had in order that we might see V-E Day. The men who were not there—the memory of them, the years we had trained with them, knowing their families, or perhaps the brief moments we had known some of them who came to us as replacements, the insight we had had into their very souls which can come only to a man who sees his soul laid bare and lives a thousand lives or dies a thousand deaths in a single day of combat.

Stretching back from the Turgen Sea to Bensheim and from Bensheim to Hamn, from Hamn to Henri Chappelle, and from Henri Chappelle to Marigny and from Marigny to St. Mere Eglise were rows of even white crosses dotted with occasional stars of David, where school children on their way home left flowers on the graves. And always flying proudly but somberly were the Stars and Stripes, symbol of the devotion of the men who rest beneath.

The men were not there. Never completely gone from our

minds were the little things we remembered—funny, crazy things they did, premonitions they had, ways they fought or talked, or maybe even things about them we hadn't liked; we supposed that we should feel our responsibility and we guessed, too, we were living on borrowed time, time loaned to us by these men who were not there. No more slopping through foxholes half filled with water, clothes damp, and with such a constant hunger for something which we could never quite place or satisfy. No more of this. No more of it for them either. We had seen the gaping wounds that had sapped their lives.

We had seen the cemeteries; we knew how they were buried. We had seen these rows of lifeless objects, shattered mockeries of that which had been breathing, pulsating men, friends of ours, buried in the soil of France and Belgium and Germany.

Strange that the soil of Germany was no different from the soil of France or America, nor were men any different. Men did what they had to do and hoped they could endure it. If they were lucky, they got back, or sometimes in the hellishness of combat they looked on the body of a soldier and said meaningly, "Won't nothing bother him anymore."

There were some other men who were not there, men in hospitals in Michigan and Washington, Atlanta and Vancouver, men whose every footstep would bear testimony to war and everything it was and everything it did to men.

And so the war was over. Perhaps we should have celebrated. Perhaps our celebration was the quiet realization that we were here, and they were not, for it was only by the grace of God, by hard fighting, and perhaps sometimes by poor shooting that we had lived to see Victory in Europe, "The shell spun close; the molten metal Tore your life away; its jagged edges Darkened eyes with suffering. I watched You die, and in that day I swore By God in heaven I would keep faith; I would remember Bullet, shell, and hell—. I saw your name today beneath a lonely cross, So white and straight within its row. There were so many of them, yet my eyes again were turned to you and God. It was not you they laid in alien sod."

21. America The Beautiful

With the cessation of hostilities, the question most frequently heard was "when are we going home?" Rumors of long occupation duty, of immediate redeployment to the Pacific theatre, and of certain shipment home were rampant.

On 12 May 1945, orders for a shift of Zone came to the regiment, and a few days later, the shift was complete, but it was only a shift of position instead of the expected pre-return movement. Regimental headquarters moved into Heilbrounn and plunged into the mass of administrative details which had accumulated. The First Battalion, Canon Company, Anti-Tank Company, and Service Company were stationed in Ansbach, a comparatively undamaged German city of some fifty thousand persons. The Hindenburg Barracks housed the troops adequately and under the rigorous cleaning details assumed the aspect of an American army camp.

The Second Battalion headquarters were in Dinkelsbuhl with the companies located nearby. The Third Battalion was at Neuen Dettelsau.

The Regiment remained in this position from the 15th of May until 9 June 1945. The three battalions set up outposts in surrounding towns and daily motorized patrols scoured the area. Bridges and important installations were guarded. Personnel not engaged in guard duty were kept busy reconditioning equipment, resuming training schedules, and recovering physically from the drain of combat.

One of the greatest problems handled by the Regiment was that of Displaced Persons, mostly Russian, French, and Polish. These people had to be processed, housed, and fed, and administrated in an orderly fashion. The largest DP Camp under the control of the regiment was located on a hill to the east, overlooking Ansbach.

German prisoners of war were processed and discharged, an enemy prison camp was maintained and carefully guarded, curfews were enforced, and careful liaison established with the Military Government in the execution of policies now established by them and carried out through the Combat Troops. This occupational zone for which the regiment was responsible included some 1,300 square kilometers.

As could be expected, life was just beginning to assume comfortable routine when movement orders were received from Division headquarters. On 10 June 1945, the regiment moved by vehicle to an assembly area near the town of Schesslitz. All units in this new area pitched pup tents or pyramidal tents.

From the 11th of June to the 21st of June, the Regiment carried out its training schedule insofar as was practical. The war against the Japanese was still raging in the Pacific, and the probability was that the 22nd Regiment would be, in due time, redeployed to that zone of action. The training was carried on accordingly. Organized athletics, personal contests, field meets, and group games were included to build up the physical condition of the troops.

Supply, now a far greater problem than during combat, was under the control of Major James A. Burnside, Regimental S-4.

Clothing and equipment were issued to the companies. Property inspections were made, and all equipment became accountable to company commanders and individuals. Unserviceable or worn out equipment was turned in and replaced. Vehicles were overhauled to the fullest possible extent. Field pieces such as anti-tank guns and 105 cannons were cleaned, repaired, and turned into Division to be disposed of through channels. Officers from the Inspector General's Department periodically inspected company and battalion supply records.

In the headquarters of all companies and battalions, as well as Regiment, painstaking care was taken to bring all the administrative details up to date. Company officers, clerks, and 1st Sergeants worked meticulously day and night to straighten out long confused records. Sick books, duty rosters and morning reports were brought up to date. Combat awards, long due men of the Regiment were sent forward, and those awards previously approved were sent to the bearer. Administration had now become an important factor that was to be taken care of, as tedious as it often seemed. I. G. inspectors also checked these records and made corrections when and where they were necessary.

For the troops, Special Service Officers arranged long awaited recreation. Men were sent on passes to rest areas, USO shows were nearby, and movies were held nightly in open air theatres on hillsides or in barns.

Parades were organized, and Divisional and Regimental Reviews took place in the surrounding fields. Awards and promotions were presented to deserving men during these reviews. In the evening, Retreat was held, and orders for that evening and the following day were issued.

The Regiment in toto, was striving to bring up to date a backlog of administrative work long encumbered by combat conditions, and which must be made current in order to efficiently carry forward the competent record held by the Regiment.

On the 22nd of June 1945, leading elements pulled out of the tent area and left by train in forty and eight box cars, forty men or eight horses as described by the French. Part of the Regiment was to move by vehicle, others by train. The vehicle convoy, made up of organic transportation, followed up a day later. The move was to be made from Bamberg, Germany, to Camp Old Gold in the vicinity of Le Havre, France. It was now apparent that the 22nd Infantry Regiment was to be redeployed to the United States through the French port. The move, both by train and vehicle, even though it was to take four or five days, presented to all an opportunity to see the beautiful countryside without the conscious fear of sudden and complete destruction.

By the 27th of June, the Regiment in its entirety had closed into Camp Old Gold. Processing began at once to prepare troops and equipment for the trans-Atlantic journey to the USA.

For almost a week, from the 28th of June until the 2nd of July, processing continued. Training periods became brief, and time was suitably used for cleaning, crating, and marking of clothes and equipment. The morale within the Regiment had risen to new heights. There were men and officers who had been overseas for more than three years; the Regiment had been overseas since January 1944. The unsurpassed combat efficiency of the units had temporarily shifted to garrison efficiency, and everybody did his utmost to prepare. Vehicles and several types of automatic weapons were turned in to Division for disposal through channels.

On **July 2, 1945**, the 3rd Battalion, less Company 'M', moved to the port of Le Havre and boarded the United States Army Transport 'James Parker'. The 3rd Battalion was to be the advance detail on board and pull the necessary details enroute. The following day, the 3rd of July, the remainder of the Regiment, less Company 'H' moved to the port and boarded the vessel. Companies 'H' and 'M' moved into the loading area and boarded the USAT 'Excelsior'.

At 1613 hours, the 'James Parker' was under way, bound for the States. During the voyage all troops were allowed to spend their time eating, sleeping, lounging about on deck in the warm sunlight, or going to the movies. Card playing and reading seemed to be prevalent among the majority of the men. The second day out of Le Havre, a slight storm was encountered, but other than this the voyage was quite pleasant.

Shortly after dawn **July 11, 1945**, the eastern coast of the United States was sighted off the port side as the transport moved northward into the Hudson River and New York Harbor. At 1000 hours the Statue of Liberty was passed, and welcoming ships encircled the 'James Parker'. At 1100 hours the army transport docked at Pier 84, New York City, N. Y. By 1700 hours, the Regiment, less Companies 'H' and 'M,' had unloaded and moved either by bus or by train to Camp Kilmer, New Jersey.

Immediately upon arrival at Camp Kilmer the processing prior to recuperation furloughs began. Within twenty four hours, the first contingent of men left Camp Kilmer by rail, bound for their nearest reception stations and their thirty day furlough papers.

Companies 'H' and 'M' embarked on the USAT 'Excelsior' the 3rd of July and shoved off about 1400 hours. After a pleasant voyage, the companies docked at Hampton Roads, Virginia, early in the afternoon of the 12th. Within twenty four hours these men were also headed home for their long deserved thirty day recuperation furloughs.

At Camp Butner, the advance detail under command of

Major Frederick T. Kent, Regimental S-2, arrived to make final arrangements pending the arrival of the main body. This detail, consisting of competent officers and enlisted men, had left Europe several days ahead of the regiment so that they could make the necessary preparations. When they arrived at the camp, they quickly established battalion and company areas, unpacked all TAT equipment, drew necessary supplies from Post Quartermaster, and in full made ready for the arrival of the Regiment.

By the 27th of August 1945, the greater part of the Regiment was assembled at Camp Butner and an intensive training program got under way. Even though hostilities had unofficially ceased in the Pacific and an inevitable armistice was close at hand, the training was to be carried on. The Regiment was scheduled to depart for the Pacific War Theatre in the early part of November 1945.

* * * * *

The 22nd Infantry Regiment, 4th Infantry Division, was now under the direction and control of the Second US Army. With the assembly of the Regiment, new problems presented themselves.

All enlisted men with eighty-five points or more were to be discharged as rapidly as possible. Officers possessing one hundred fifteen points, or more, were also eligible for discharge at this time if they desired. At once discharging began. Training was initiated with special emphasis placed on military courtesy, personal hygiene, athletics, and physical conditioning.

On 2 September 1945, word was received of the official surrender of all Japanese forces, land, sea, and air. This sudden cessation of hostilities would certainly cause wide changes in plans, but temporarily the action of the 22nd Infantry remained the same. Training was broadened to include weapons firing and qualification of all personnel in their individual arms and crew served weapons. About the middle of the month, orders were officially sent down that the Regiment would not go to the Pacific as previously planned but would stay in strategic reserve for a period of time then unknown.

The Army's Adjusted Service Rating, point score, dropped for both officers and enlisted men the 1st of October and this placed virtually hundreds of men eligible for immediate release from the service. The change also meant that there was to be quite a turnover of personnel within the Regiment. Shortly after the 1st, a directive came down from Army Headquarters authorizing forty-five day furloughs for men not immediately due for discharge, provided that forty-five percent of TO strength remained for duty. With these two changes in effect, training became secondary, and administration became the top issue.

The remainder of October was spent in complying with the new discharge scores, the new furlough policy, and the indoctrination of new replacements being received almost daily. Parades and reviews were conducted throughout. Awards and decorations were made during such parades to the men who had been awarded them. Each of the nineteen companies organized information and education rooms for the use and benefit of all personnel.

With the coming of November, point scores for officers and enlisted men were again reduced, making another large number of men eligible for discharge. As before, administration took precedence over any other type of work, Company records were continuously being inspected, and new directives in regard to these records were being complied with. The 22nd Infantry was shifting from a Combat Team to a garrison Regiment.

During the month, the forty-five day furlough policy was stopped. The Regiment, still a part of the 4th Infantry Division, was shifted to the command of the First Army under the command of Lieutenant General Courtney H. Hodges. Again training schedules were formulated and carried out to include all types of training and education. New weapons were introduced. Schools of various types were made available to both officers and enlisted men. Quotas were received and filled for these schools in an effort to bring the latest ideas within the Regiment.

22. Fourragere

V-J Day probably made the men of the Fourth Infantry Division as happy as any of the troops anywhere for the reason that it had been slated for an assault landing on the beaches of Japan. Toward this end, the division had been ordered to Camp Butner, North Carolina, immediately following a thirty day period of rest and recuperation.

It was a jubilant regiment that reassembled at Camp Butner; jubilant because the end of the War meant almost certainly they would not now go overseas.

It was amazing, this change of the fighting soldier to the garrison soldier. Within a matter of weeks, high point men began to be discharged. Soon some of the men who had contributed most to its success and ability were gone. Joe Samuels, George Goforth, Swede Henley, Hanshaw, Sloninsky, Carol Kemp, Tom Harrison, Magruder, and hundreds of others soon took off their khaki and returned to as many walks of life.

These were lonely days, days in which comrades with whom we had lived and fought were separated from us. It was not an easy thing, this parting.

In the months which followed, the regiment changed personnel rapidly. Since our own men were being discharged, it was necessary to fill the complement with men from another source, although the Twenty-Second had many re-enlistments, chiefly because of the spirit of its officers and men. It received many fine officers and men from the 87th and 95th Divisions, which had been inactivated. Others came from the 76th, the 100th, and the 99th Divisions.

Throughout the winter months, routine training, orientation classes, and administration were carried on only with extreme difficulty, due to the shortage of trained personnel.

On January 2, 1946, training was resumed following the Christmas Holidays. It was shortly thereafter that rumors reached the regiment that the Division might be inactivated. It seemed incredible to the men of the regiment, and no one believed that the War Department would seriously consider inactivating a Regular Army Division which was made up of three of the oldest Regiments in the Army. It soon became apparent that it was not rumor but fact, and the orders were received the first week of February with instructions that the inactivation would be completed by 1 March 1946. No reason was given.

In one terse order, the War Department had accomplished what an entire enemy army could not. The order, which was to inactivate the Fourth Infantry Division, Twenty-Second United States Infantry Regiment, and the other division units, had to be obeyed. Perhaps all this was a necessary and a fitting rest for a gallant regiment which had turned its regimental motto, "Deeds Not Words, " into an accolade which few regiments could equal and none excel.

The days of February and March were unhappy, busy days in which the question of individual future often came to mind. Whether or not to stay in the Army, whether or not to accept a regular army commission, opportunities for advancement in return to civilian life, and separation from tried and true friends were uppermost in the minds of the regiment.

On the 20th of February 1946, Colonel John F. Ruggles, who had commanded from Sellerich, Germany, took a leave prior to his assignment to the Command and General Staff School at Fort Leavenworth. This placed the Regiment in the hands of Lt. Colonel Arthur S. Teague. Great soldier, excellent tactician, leader of men, and loyal friend of both his officers and men, Art Teague was loved and respected by the Twenty-Second Infantry Regiment as were few men. He had served as an officer with the regiment almost five years. Thus, it was fitting that he should inactivate the regiment, and the men were glad that he was thus honored.

At the last formal review on the parade ground at Camp Butner, North Carolina, all sensed this was an historic occasion and certainly the last of its kind for the famous Fourth in World War II. General Courtney H. Hodges, Commanding General of the First Army, the Belgium Ambassador to the United States, a representative of the French Embassy, Major General Raymond O. Barton, Major General Harold W. Blakely, and Major General Melborn were guests of honor and present in the stands for the review.

The Regiments formed and moved on to the parade ground, in order, the 22nd, the 12th and the 8th, followed by the Field Artillery, Engineers, and the special troop units. The day was gloomy and there was a slight drizzle as if Nature herself were weeping to watch such an organization die. There was grumbling in the ranks among the new men who had no loyalty to the Division, but there was a stillness that mirrored an ache in the hearts of the old Fourth Infantry Division men.

Colonel Ruggles, senior Regimental Commanding Officer in the Division, was commander of troops.

Combat streamers were awarded to the Colors of the Regiment, and there was a thrill of pride within the heart of each man as the Regimental Colors dipped to receive the ribbons. And then the Belgium Ambassador was introduced. He was a tall, stately man who carried the dignity and the honor of brave Belgium upon his own shoulders. Scarcely had he started to speak when death-like stillness fell over the entire assembled Division. It seemed as if each soldier sensed that here was something he wanted to hear.

The Ambassador spoke, "Belgium learned to love and honor the Fourth Infantry Division in the first World War when on the banks of the Marne the blood of your men mingled with the blood of our own, and the fierce Huns were stopped. Again in this war it was fitting that the Fourth Infantry Division should play so large a part in the liberation of Belgium, who had suffered so much at the hands of a cruel and ruthless enemy. We knew that you would come, and in that knowledge, liberty still lived within our hearts.

"Belgium salutes the brave men of the Fourth Infantry Division. She salutes Lt. Colonel Mabry, General Roosevelt, and Sgt. Macario Garcia (Medal of Honor recipients). My country has conferred upon the men of this Division the highest honor it is in her power to bestow, and in honoring you she honors herself. The red of the Fourragere is for the blood of your men shed for the liberty and for the freedom of Belgium. The green is for the constant memory of these men and what they did, and so the Fourth United States Infantry Division will always live in the heart of Belgium. Vive la America!"

No one stirred. Somehow it was fitting. Somehow it was appropriate that such an honor should come to the battle weary, exhausted, broken hearted, proud Division, and to her great fighting Regiments. Then came the order, "Pass in review." The men marched stiffly and well even in the mud and drizzle, and as the colors passed by, every person snapped to attention. As each one realized that this was the last time he would march as a member of the Double Deucers and of the famous Fourth Infantry Division, there was a stillness and heartache which can be occasioned only by the death of a well-beloved friend.

"Eyes right." Heads snapped. The generals looked at the soldiers. The soldiers looked at the generals. Neither saw the other, but rather the foxholes and hedgerows of Normandy, the crosses at St. Mere Eglise and Henri Chappelle, the match sticks and mud of Hurtgen. They saw marching in ranks, in file after file with perfect cadence and deathless spirit, all of the men who were not there. Not there? Certainly they were in the hearts and minds of those who remember, never to forget, in the love of those who would never cease missing them, in the freedom of every American.

* * * * *

And so the men marched off of the parade ground and into the cities, villages, farms, offices, or other army posts. And with them went the Twenty-Second United States Infantry Regiment. A dead regiment? Certainly not. Not so long as a single man still lives and remembers. Sleeping, perhaps, but not dead. The Twenty-Second United States Infantry Regiment, the finest Regiment, the beloved Regiment, our Regiment, which gave life to the motto forever etched in our hearts, "Deeds Not Words."

Addenda — Citations

HEADQUARTERS
4th INFANTRY DIVISION
APO 4, US ARMY •
15 July 1944

ORDER OF THE DAY:
NUMBER 1:

COMMENDATION FOR MERITORIOUS SERVICE
3rd Battalion, 22nd Infantry

On 10 July 1944, the 3rd Battalion of the 22nd Infantry was committed against the enemy. On 11 and 12 July 1944, the attack was resumed with the 3rd Battalion advancing to seize an objective in the vicinity of PAIDS by envelopment from the east. The 22nd Infantry less the 3rd Battalion was relieved by elements of the 12th Infantry on the night of 12-13 July 1944. The 3rd Battalion 22nd Infantry was attached to the 12th Infantry and remained in constant contact with the enemy until 1800, 14 July 1944.

During this period the Battalion was constantly fighting heavy tanks and infantry. It was subjected to enemy small arms, mortar, and artillery fire.

After being engaged for five days, the 3rd Battalion 22nd Infantry was still prepared to further their position and to push the attack when ordered. Each officer and man of the Battalion is to be commended for his complete devotion to duty while engaged with the enemy.

By command of Major General BARTON:
R. S. MARR, Lt. Col. GSC Chief of Staff.

* * * * *

HEADQUARTERS
4th INFANTRY DIVISION
APO 4, US ARMY •

ORDER OF THE DAY: 3 August 1944

NUMBER 10:
COMMENDATION FOR MERITORIOUS SERVICE
22nd Infantry—44th FA Bn—Co C. 4th Med Bn—1st Plat.
Co. C. 4th Engr C Bn

At 261100B July 1944, CC (A) commanded by Brigadier General Maurice Rose of the 2nd Armored Division, with the 22nd RCT as an integral attachment, jumped off on an attack which was a part of the initial breakthrough of the German lines.

The 1st and 2nd Battalions of the 22nd Infantry, with normal combat attachments, roared to the attack riding tanks. The 3rd Battalion, in reserve, fought as armored infantry.

The lightening-like thrust spearheaded by the 22nd RCT attack north of CANISY, in the vicinity of ST. GILLES, and by dark had their advance elements on their objective which was the high ground surrounding LE MESNIL HERMAN. By early morning of 27 July the objective was taken, and all enemy resistance had been eliminated.

The officers and men of the 22nd RCT are commended for the successful execution of their mission, which was accomplished with a minimum loss of personnel and equipment.

By command of Major General BARTON:
R. S. MARR, Colonel, GSC, Chief of Staff

* * * * *

HEADQUARTERS COMBAT TEAM 22 A. P. O. #4, US Army

18 August 1944
(The following letters will be read to all members of this command prior to 201800 Aug 1944)

SUBJECT: Commendation.
TO: All members of Combat Team 22.

1. It is with pleasure that I publish the following letter of commendation from Brig. General MAURICE ROSE and the accompanying endorsements from the Commanding General, 2d Armored Division and the Commanding General of our own 4th Infantry Division.

2. The breakthrough, spearheaded by Combat Team 22 and the 66th Armored Regiment grouped as Combat Command "A", made history. It will be studied by professional soldiers for generations to come. The magnificent battle you fought in this operation will rank with your equally magnificent landing on D-Day.

The story of your achievements will endure as long as military history endures.

3. To every officer and man of Combat Team 22 goes my admiration and my affection. In the great operations and hard battle still before us, may we all remember that in our hands rests the honor of a great Combat Team that knows only one direction of movement—FORWARD!

(s) C. T. LANHAM Colonel, 22nd Infantry, Commanding.

* * * * *

"HEADQUARTERS CC "A" 2d Armored Division
A. P. O. 252 c/o Postmaster, U. S. Army 2 August 1944
Commendation.
Commanding General, 4th Infantry Division.
(Through, Commanding General, 2nd Armored Division A. P. O. 252, U. S. Army).

1. I wish to commend the officers and men of the 22nd RCT, 4th Infantry Division, for their outstanding performance of duty and fighting qualities during a recent operation.

2. This organization reported to me as an infantry organization of the normal infantry-type division which, after a short period of indoctrination, operated as armored infantry in a manner that would do credit to any regular armored infantry regiment in the army. They perfected a teamwork between tanks and infantry that very nearly approached perfection. This, together with their fighting qualities, made them an irresistible force on the battlefield.

3. The professional qualities, adaptability, and far-sightedness

of the commander of this organization, Colonel C. T. Lanham, 0-15568, was a motivating force behind the welding of this infantry regiment into an armored combat team.

4. I shall always look with pride to the time when I was in command of an organization that included the 22nd RCT.

/s/ Maurice Rose /t/ MAURICE ROSE Brig. Gen., U. S. Army

* * * * *

(2 Aug 44) 1st Ind. /djb
HEADQUARTERS 2D ARMORED DIVISION, A. P. O. #252. 3 August, 1944.
TO: Commanding General, 4th Infantry Division, A. P. O. #4.
Forwarded with pleasure and with my personal commendations for an outstanding job, well done.
/s/ Edward H. Brooks, /t/ EDWARD H. BROOKS, Major General, U. S. Army Commanding.
AG 201.22
(2 Aug 44) 2d Ind.
HQ 4TH INF DIV., APO 4, US Army, 12 August 1944. To:, CO, 22d RCT, APO 4, US Army.

1. It is heartening indeed for your Division Commander to receive such an outstanding report of your exemplary actions during the recent period when you were attached to an armored division.

2. The gallantry, ability and initiative as displayed by the officers and men of CT 22 reflects great credit not only on each of you as individuals but your organization and the 4th Infantry Division.

3. It is with extreme pleasure and approbation that I add my express of appreciation and praise to this commendation.

/s/ R. O. Barton, /t/ R. O. BARTON, Major General, U. S. Army, Commanding.

* * * * *

HEADQUARTERS 4TH INFANTRY DIVISION APO 4, US ARMY
20 August 1944
ORDER OF THE DAY:

NUMBER 22:
COMMENDATION FOR MERITORIOUS SERVICE 22ND REGIMENTAL COMBAT TEAM

On the evening of 10 August 1944, it became necessary to move a reinforced RCT of the 4th Infantry Division from the ST POIS area to the vicinity of LE TEILIEUL and PASSAIS, a distance of some thirty-five miles. This mission was given to the 22nd RCT.

The Division Commander issued a brief movement order to the Commanding Officer of the 22nd RCT at 1420 hours. Attached to this RCT were Company C, 70th Tank Battalion, 4th Reconnaissance Troop, (Mech), Company B, 801st TD Battalion, one platoon, Company C, 634th TD (SP) Battalion, and sufficient 2-1/2 ton trucks from the 4th Quartermaster Company to motorize the RCT. The situation in the sector in which this RCT was to move was obscure and little was known of the situation of the enemy.

The Commanding Officer of the 22nd RCT returned to his

command post at approximately 1445 hours and his subordinate commanders began reporting at 1500 hours. As each subordinate commander arrived, he was issued a fragmentary oral order and dispatched. Leaders of some of the attached units were unable to arrive prior to 1530 hours. Throughout the period during which these orders were issued, the Command Post was subjected to heavy enemy shelling, including three hits from shells of heavy caliber, which resulted in disrupted communications and many casualties.

Despite the rapid oral orders and the adverse circumstances under which they were issued, the 3rd Battalion, 22nd Infantry, passed the RCT IP in ST POIS at 1700 hours, a distance of some five miles from its assembly area. Two miles further on it effected a rendezvous with Company C, of the 70th Tank Battalion, mounted its leading company on tanks, and awaited the passage of the 4th Reconnaissance Troop (Mech) at this point. At 1740 hours the Reconnaissance troop cleared and the entire RCT followed.

All rendezvous of subordinate elements and attached units were affected rapidly and with precision. The thirty-five mile move to the new area was made smoothly and without accident or casualty. The excellence of this move is further enhanced by the fact that it was ordered at a time when the RCT was not on an alert basis and when two of its three battalions were operating in extensive outposts.

The officers and men of the 22nd RCT are hereby commended for the highly successful accomplishment of an extremely difficult mission and for outstanding performance of duty.

By command of Major General BARTON:
R. S. MARR, Colonel, GSC Chief of Staff

* * * * *

HEADQUARTERS
4th INFANTRY DIVISION
APO 4, US ARMY •
ORDER OF THE DAY: 27 August 1944

NUMBER 22:
SECTION II:
COMMENDATION FOR MERITORIOUS SERVICE
22ND REGIMENTAL COMBAT TEAM

The 22nd Regimental Combat Team comprised of the 22nd Infantry Regiment, 44th Field Artillery Battalion, Company C 4th Medical Battalion, 1st Platoon Company C 4th Engineers Combat Battalion, Company C 70th Tank Battalion, and Company C 893 Tank Destroyer (SP) Battalion, was alerted at approximately 2200 24 August 1944 and given the mission of securing a bridgehead across the SEINE RIVER south of PARIS in the vicinity of CORBEIL.

The 3rd Battalion, 22nd Infantry, supported by elements of the 893rd Tank Destroyer (SP) Battalion, was given the mission of taking the initial crossing in the vicinity of CORBEIL. By daylight on the morning of 25 August 1944, Companies K and L were on the west river bank prepared to make the initial attempt and did so at 0645. Only four rubber boats were subjected to heavy enemy machine gun and AA fire which sank all four rafts in midstream and caused 17 casualties. In the meantime, the 2nd Battalion, 22nd Infantry, had moved into position along the river bank approximately 2500 yards north of the 3rd Battalion, and there attempted a crossing on an improvised raft, as no rubber boats were available. However, due to heavy enemy fire, they were forced to withdraw. At approximately 1400 hours 25 August 1944, the 2nd Battalion again attempted a crossing on the same improvised raft they had used previously, and under extreme difficulties, successfully made the crossing. By shuttling, the 2nd Battalion succeeded in landing 25 men on the east bank of the river. At about 1500 hours, rubber boats were obtained, and the entire battalion successfully made the crossing, thus securing a bridgehead on the east bank of the SEINE RIVER.

During this period the 1st Battalion, 22nd Infantry, moved to the river between the 2nd and 3rd Battalions and succeeded in getting one company across the river. This company secured the left flank of the 2nd Battalion and also secured the site on the east bank where Company C 4th Engineer Combat Battalion later succeeded in bridging the river. The 3rd Battalion then succeeded in getting one company across the river at the same spot used by the 2nd Battalion, thus securing the right flank of the 2nd Battalion.

The bridge site having thus been secured, Company C, 4th Engineer Combat Battalion was enabled to begin construction of the bridge at 0600 26 August 1944 and the bridge was completed and open to traffic at approximately 1200 hours of that day.

The Division Commander takes extreme pleasure in hereby commending the members of the 22nd Regimental Combat Team for the highly successful accomplishment of a most important mission.

By Command of Major General BARTON:
Richard S. Marr, Colonel General Staff Corps, Chief of Staff.

* * * * *

HEADQUARTERS
4th INFANTRY DIVISION
APO 4, US ARMY •

ORDER OF THE DAY)
NUMBER 37) 5 October 1944

COMMENDATION FOR MERITORIOUS SERVICE
REGIMENTAL AND BATTALION COMMUNICA-
TIONS PLATOONS, 22ND INFANTRY

From D-Day to the termination of the Cherbourg Peninsular Campaign, over seven hundred (700) miles of W-110 wire was competently laid within the 22nd Infantry's zone of operation. The task of keeping this wire intact, while it was exposed to incessant mortar and artillery fire, was a great one. Day and night the wiremen worked the lines, often in rough areas infested with enemy snipers and patrols.

In the vicinity of DOMFORT, FRANCE, while containing the enemy, over seventy-five (75) miles of W-110 wire was laid to all positions. This wire was constantly patrolled and maintained, the wire crews working without rest for period of thirty-six (36) hours and more. In spite of this prolonged effort, maximum efficiency was maintained.

When the hedgerows were left behind, the 22nd Infantry was on the move over great distances for long periods of time. For communication, the regiment was entirely dependent on radio. Radio operators and repairmen provided and maintained this communication with ceaseless and tireless effort. Their efforts contributed in a large measure to the overall success of these follow-up operations.

This outstanding performance, during the period 6 June 1944 to date of this order, was not without combat losses. Two communications officers were killed, one seriously wounded, and many of the enlisted personnel were lost to the platoons. The officers and men who took their places picked up their work and carried on with the same aggressiveness spirit and devotion to duty.

The division commander desires to congratulate and commend each member of the regimental and battalion communication platoons of the 22nd Infantry for an outstanding performance of duty over a prolonged period of difficult military operations. The spirit of cooperation and unselfish devotion to duty displayed therein has enabled the 22nd Infantry to accomplish mission after mission along the road to victory and is highly worthy of emulation.

By Command of Major General BARTON:
RICHARD S. MARR, Colonel, General Staff Corps., Chief of Staff.

* * * *

HEADQUARTERS
4th INFANTRY DIVISION
APO 4, US ARMY •

ORDER OF THE DAY:
23 October 1944
NUMBER 40
COMMENDATION FOR MERITORIOUS SERVICE

22ND COMBAT TEAM PATROLS

The following is quoted from a letter, Headquarters 22nd Infantry, subject, "Recommendation for Order of the Day", date 19 October 1944, received by the Division Commander:

"On the evening of 8 October 1944, the commanding officer of Combat Team 22 was planning to make an attack on the Siegfried Line in the vicinity of Udenbreth, Germany. Certain information concerning tank obstacles, minefields, avenues of approach, and relative terrain features was lacking. This information could only be determined by vigorous reconnaissance patrols. Seven patrols given the mission of securing this information the previous night were unsuccessful due to minefields, heavy enemy patrols, and heavy small arms fire. The commanding officer of Combat Team 22 then directed that further patrols be assigned the mission of securing the desired information. Two patrols were formed. One patrol was under the leadership of 1st Lieutenant John A. Fischer, 01315487. The members of this patrol were Staff Sergeant James J. Hearnee, 34582154, Sergeant James A. Colvin, 34582203, Private First Class Willie D. Jackson, 34889040, and Private First Class Roy R. Jasper, 35809355. The other patrol was under the leadership of Sergeant Attilio DeLorenzo, 33088197. The other members of this patrol were Private First Class James A. Patterson, 35706800, Private First Class Herbie E. Hebert, 38267686, and Private Joseph E. Bessette, 11062236.

The route assigned these patrols was over difficult terrain, going approximately 1,000 yards behind the enemy lines. The enemy was known to have numerous anti-personnel mines, barbed wire, and heavy outposts with machine gun crossfire. Before leaving on this mission, Lieutenant Fisher and Sergeant DeLorenzo and their men made careful and thorough study of the aerial photos and maps of the routes they were going to follow. The patrols started on their missions at dusk.

Soon after reaching the 1st Battalion outpost line, an enemy patrol was seen; this was successfully by-passed. As the patrols proceeded, enemy outposts fired on them, but these, too, were by-passed. Continuing on their mission, the patrols were fired on several times, but the determination to secure the necessary information spurred the men on.

Upon reaching their objectives, a careful study of the terrain and obstacles was made; the stream was checked for approaches, width, and depth; the dragon's teeth were checked for location, construction, width, and number; trees were checked for size and density. All necessary information for an attack in this sector was obtained. The patrols returned to their battalion areas at 2300 hours.

The performance of these patrols is commendable. Their courage, resourcefulness, and perseverance reflect credit on themselves, their regiment, and their division.

The patrol leaders and members of the patrols are commended for the successful accomplishments of a hazardous mission and for an outstanding performance of duty."

It is with extreme pleasure and approbation that the division commander adds his commendation for such an outstanding performance of a hazardous and difficult mission.

(Order of the Day No. 40, Hq 4th Inf. Div., 23 Oct 44)

By command of Major General BARTON:
Richard A. Marr, Colonel, General Staff Corps, Chief of Staff.

* * * * *

GENERAL ORDERS #14:

IV BATTLE HONORS. -1. As authorized by Executive Order No. 9396 (sec. I, "Bui. 22, WD, 1943), superseding Executive Order No. 9075 (sec. Ill, Bui. 11, WD, 1942), the following units are cited by the War Department under the provisions of section IV, Circular No. 333, War Department, 1943, in the name of the President of the United States as public evidence of deserved honor and distinction. The citation reads as follows:

The 22nd Infantry Regiment is cited for extraordinary heroism and outstanding performance of duty in action in Normandy, France, during the period 26 July to 1 August 1944. The 22nd Infantry Regiment was the infantry element of an armored-infantry combat command which successfully effected a breakthrough of the German line of resistance west of St. Lo, forming the St. Gillis-Marigny gap through which armored-infantry columns surged deep into German-held territory. Operating against hardened infantry, artillery, and panzer units, this regiment, often riding its accompanying tanks, met and overcame the stiffest German resistance in desperate engagements at St. Gillis, Canisy, Le Mesnil, Herman, Villebaudon, Moyen, Percy, and Tessysur-Vire. The 22nd Infantry Regiment, in its first action with an armored division, after a short period of indoctrination, assumed the role of armored infantry with unparalleled success. Throughout the swiftly-moving, 7-day operation, the infantry teams kept pace with the tanks, only resting briefly at night relentlessly to press the attack at dawn. Rear echelons fought with enemy groups bypassed in the assault. There was little protection from the heavy artillery which the Germans brought to bear on the American armor. Enemy bombers continually harassed the American troops at night, but in an outstanding performance of duty, the 22nd Infantry Regiment perfected an infantry-tank team which, by the power of its determined fighting spirit, became an irresistible force on the battlefield.

BATTLE HONORS

Distinguished Unit Citation 3rd Battalion 22nd Infantry:
For D-Day assault: GO#26 HQ FUSA 17 June 44

Distinguished Unit Citation 22nd Infantry Regiment:
For St, Lo Breakthrough 26 July 44 to 1 August 44: Sec. 4 GO#14 WD 3 March 45.

Regimental Commanders

22ND INFANTRY REGIMENT COMMANDERS FROM NOVEMBER 1939 TO MARCH 1946

Colonel Albert S. Peake
November 10, 1939 — November 11, 1941

Colonel George H. Weems
November 12, 1941 — February 17, 1942

Colonel Hervey A. Tribolet
February 18, 1942 — June 10, 1944

Colonel Robert T. Foster
June 11, 1944 — July 8, 1944

Colonel Charles T. Lanham
July 9, 1944 — March 2, 1945

Colonel John F. Ruggles.
March 3, 1945 — February 19, 1946

Colonel Arthur S. Teague
February 20, 1946 — March 1946

Ct-22 Command Posts

(Note: Vic means 'In Vicinity of')

6 June 1944	St, Martin-de-Varreville, France
7 to 9 June	Ravenoville
9 to 12 June	Azeville
13 to 17 June	St. Floxel
18 to 22 June	Vic Le Theil
23 to 30 June	Genneville
1 to 5 July 1944	Amfreville
6 to 7 July	Vic Carentan
8 to 14 July	Vic Sainteny
15 to 16 July	Montmartin-en-Graignes
17 July	St. Jean De Daye
18 to 25 July	Vic Le Molay
26 July	Pont-Hebert
27 July	Canisy
28 to 29 July	Le Mesnil Herman
30 July	La Denisiere
31 July to 1 August 1944	Villebaudon
2 August	Hambye
3 August	Villedieu-Les-Poeles
4 to 7 August	St. Pois

8 to 9 August	Chateau Lingeard
10 to 16 August	Le Teilleul
17 to 22 August	Carrouges
23 to 24 August	Ablis
25 to 27 August	Corbeil
28 August	Paris
29 August	Les Mesnil Amelot
30 August	Ermenonville
31 August	Chateau Vez
1 September 1944	Crecy
2 September	Landrecies
3 to 6 September	Pommereuil
7 September	Graide, Belgium
8 to 9 September	Smuidd
10 September	Heuffalize
11 Septembe	Beho
12 September	Hemmeres, Germany
13 September	Schweiler
14 September to **4 October 1944**	Buchet
4 to 7 October	Honsfeld, Belgium
7 to 23 October	Murringen
23 to 31 October	Krinkelt
31 October to **8 November 194**	Krinkelt
9 to 19 November	Zweifall, Germany
19 November to 3 December 1944	Vic Grosshau
3 to 4 December	Vic Zweifall
5 to 10 December	Senningen, Luxembourg

10 to 21 December	Mondorf
21 December to **7 January 1945**	Rodenbourg
7 to 17 January	Boudelet
17 to 28 January	Ferme Pletschette
28 January to **3 February 1945**	Hautbellain, Belgium
3 to 6 February	Vic of Bleialf, Germany
6 to 8 February	Buchet
8 February to **2 March, 1945**	Sellerich
2 to 4 March	Tafel
4 to 5 March	Dausfield
5 to 6 March	Schwirzheim
6 to 8 March	Scheuern
8 to 14 March	Oberbettingen
14 to 20 March	Magnieres, France
20 to 26 March	Hochstett
26 to 30 March	Maschine, Germany
30 March to **1 April, 1945**	Hebstahl
1 to 3 April	Paimar
3 to 5 April	Grunsfeld
5 to 7 April	Marbach
7 to 10 April	Bad Mergentheim
10 to 11 April	Herbsthausen
11 to 13 April	Weikersheim
13 to 14 April	Laudenbach
14 to 16 April	Wildentierbach
16 to 18 April	Spielbach
18 to 20 April	Haufelwinden

20 April	Lunbach
21 April to 22 April	Gronongen
22 April	Altenmunster
22 to 23 April	Hummelsweiler
23 April	Rosenberg
23 to 24 April	Dewangen
24 to 25 April	Essingen
25 to 26 April	Brenz
27 to 28 April	Gundremmingen
28 April	Aretsried
28 to 29 April	Graben
29 to 30 April	Jesenwang
30 April to **1 May, 1945**	Gauting
1 to 2 May	Hoen Schaftlarn
2 to 3 May	Behrloh
3 to 4 May	Thalam
4 to 15 May	Schwabach
15 to 31 May	Heilsbronn

Casualties 22nd Infantry Regiment

ENLISTED MEN / OFFICERS

June — 1944
Wounded 1560 / 104   Killed 373 / 23

July — 1944
Wounded 1044 / 56   Killed 263 / 16

August — 1944
Wounded 381 / 26   Killed 65 / 7

September — 1944
Wounded 433 / 31   Killed 93 / 11

October — 1944
Wounded 102 / 9   Killed 15 / 1

November — 1944
Wounded 1641 / 89   Killed 208 / 25

December — 1944
Wounded 570 / 21   Killed 186 / 3

January—1945
Wounded 89 / 3   Killed 18 / 1

February—1945
Wounded 728 / 44   Killed 136 / 10

March—1945
Wounded 361 / 14   Killed 101 / 2

April—1945
Wounded 376 / 21   Killed 88 / 7

May—1945
Wounded 2 / 1   Killed 1 / 0

TOTAL
Wounded 7,287 / 419   Killed 1,547 / 106

TOTAL CASUALTIES—Wounded and Killed
    ENLISTED 8,834
    OFFICERS   525   COMBINED = 9,359

17.5% of Total Enlisted Casualties were killed.

20.2% of Total Officer Casualties were killed.

These figures are a compilation of the most accurate sources available through the S-1 Section, 22nd Inf.

Awards And Decorations

The 22nd Infantry Regiment Officers, NCOs, and Enlisted Men received the following awards and decorations during the period 6 June 1944 to 8 May 1945:

Medal of Honor	1
Distinguished Service Cross	15
Silver Star	330
Legion of Merit	2
Soldier's Medal	1
Bronze Star Medal	1,692
Bronze Star Medal (Meritorious Service)	181
Foreign Awards	25

These totals are still incomplete as orders are still being printed awarding individuals.

(This statement is as of Chaplain Boice's finalizing his book in 1959).

About the Authors

William Speas "Bill" Boice was born in Blackfoot, Bingham County, Idaho on December 21, 1915, the second of two children born to William A. Boice and Geneva Speas Boice.

In 1922, he and his family moved to Maricopa County, Arizona, where, by 1933, they had permanently established themselves in Phoenix.

Bill graduated from Union High School in Phoenix in 1935. He attended the University of Cincinnati, where he received a Bachelor of Science in Law degree in 1939, and a Bachelor of Arts degree in 1943.

He was commissioned a First Lieutenant and Chaplain (Army of the United States) on June 14, 1943. He entered active duty with the Army on July 11, 1943, as a Chaplain with the 22nd Infantry Regiment of the 4th Infantry Division. He sailed with the Regiment to England aboard the *Capetown Castle* on January 18, 1944. On April 1, 1944 he was promoted to Captain.

Bill served with the 22nd Infantry through all five of its campaigns in Europe, and was awarded the Silver Star, and the Bronze Star Medal with oak leaf cluster. In 1945/1946 he was the Regimental Chaplain of the 22nd Infantry at Camp Butner, North Carolina. When the Regiment was inactivated in early

1946, Bill remained on active duty in the Army, and was promoted to Major (AUS) on August 30, 1946.

In 1945, on the ship to the United States after World War II in Europe ended, Bill, along with then Colonel John F. Ruggles and several other officers of the 22nd Infantry, formed The Twenty Second Infantry Officer's Association. The Association eventually became the 22nd Infantry Regiment Society, and is still active today, with veterans of all ranks and eras as members.

On June 19, 1947, Bill was given a commission as a Captain in the Regular Army Chaplain Corps, with date of rank back to December 21, 1943. In 1948, he received the Army Commendation Medal. In 1949, he was the Post Chaplain at Fort Meade, Maryland. He was Wing Chaplain at Perrin Air Force Base, Texas in 1951. He was separated from active duty in 1951.

Bill became the founding minister of First Christian Church in Phoenix, Arizona in 1952. The First Christian Church building was designed by the famous architect Frank Lloyd Wright and is considered one of the ten best church buildings ever built in America by the Church Architects Guild of America.

He became Minister Emeritus of First Christian Church in 1984. In addition to his ministry, Bill was the Religious News Commentator for Radio KOY in Phoenix for 24 years.

His education included graduation from Cincinnati Bible Seminary, Cincinnati Conservatory of Music, and Harvard Chaplain's School. He received Honorary Degrees from Southwest Christian Seminary—Doctor of Divinity; Kentucky Christian Bible College—Doctor of Theology; and Pacific Christian College—Doctor of Divinity.

William S. Boice was never married and died at the age of 88 on December 23, 2003. He is buried in Greenwood Memory Lawn Cemetery, Phoenix, Maricopa County, Arizona.

As noted in the front of this book, Dr. Boice gave strong credit to Lt. David James who did the After Action Report research upon return to the U.S. at the end of WWII. His bio follows:

David Randolph "D.R." James was born in El Dorado, Union County, Arkansas on September 22, 1924, the son of Randolph and Georgia Hughes James. David graduated from Gulf Coast Military Academy in Gulfport Mississippi in 1943, then entered the Army under Selective Service on June 15, 1943 at Little Rock, Arkansas. He attended Officer's Candidate School at Fort Benning, Georgia, and was commissioned a 2nd Lieutenant of Infantry on November 9, 1943.

He served at Fort Sill, Oklahoma and then deployed overseas in 1944 aboard the *SS Ile de France* as a replacement officer. He was assigned to Company M 22nd Infantry 4th Infantry Division from the 92nd Replacement Battalion on November 23, 1944, while it was engaged in the intense Battle of the Hürtgen Forest. He commanded the heavy mortar platoon for Company M 22nd Infantry in a fierce action against a German counterattack on December 3, 1944, in which his platoon of six 81mm mortars fired 1100 rounds at the enemy in thirty minutes. He saw further action in the Battle of the Bulge and the battle for the city of Prüm, Germany.

David was promoted to 1st Lieutenant on February 17, 1945, and the next day was assigned as Executive Officer of Company M. On May 17, 1945 he was transferred to Company K as a Platoon Leader. He returned with the Regiment to the United States in July 1945 and was with Company K at Camp Butner, North Carolina. He was discharged from the Army as a Captain in May 1946.

David married Dorothy Jean Callaway on June 12, 1947. He and his wife eventually had two sons. He graduated from the University of Arkansas and began a career in the hotel business. From 1949 to 1962 he managed the Randolph Hotel in El Dorado, Arkansas, a hotel built and run by his grandfather and father. He became an early Holiday Inn franchise owner, and built and managed Holiday Inns in Arkansas, Texas, and Louisiana.

He was active in civic and charitable organizations and in the First United Methodist Church. David R. James died at the age of 91 on February 24, 2016, and was buried with full military honors in Arlington Memorial Park, El Dorado, Union County, Arkansas. His awards include the Silver Star, Bronze Star Medal with oak leaf cluster, and the Purple Heart.

www.ingramcontent.com/pod-product-compliance
Lightning Source LLC
Chambersburg PA
CBHW022101210326
41518CB00039B/355